The Moon of Venus and the Investigations into the Earlier Observations of this Moon

Ferdinand Schorr

Translated from German by G. R. Stempien

G. Stempien Publishing Company
Newport, Cymru (of Wales)

ISBN 978-0-930472-90-0

TRANSLATOR'S NOTE

As an amateur astronomer, I had planned on writing a much needed book about the attempt to discover the moon of Venus. One of the primary sources was the book that is being translated here from the original German which was written by Dr. Ferdinand Schorr. However, Dr. Schorr's work was of such a high quality it was determined that nothing could be added to his fine essay on the topic that could possibly improve it. Thus, creation of a new edition was abandoned in favor of translation of Schorr's work.

All of the major observations of the phantom moon of Venus occurred exclusively during the 17th and early 18th centuries, and then came no more. Many people claimed that this was because there was no moon to see in the first place. They argued that the "phony" moon viewed by the most eminent astronomers of the day was simply the result of telescopic tricks of lighting.

This suggestion is beyond absurd; it is insulting to some of the greatest astronomers who ever lived: Cassini, Gruitihuisen, Schroeter and many others. These were professionals who were not easily deceived and, because they were professionals, they in fact checked against such a possibility of telescopic error and found none. In fact, they amassed such a supremely impressive number of verified sightings of the fleeting moon that a file of data of this vast amount would normally satisfy the claims of the discovery of any other new object in space.

This data also disproves the theories that observers were seeing Uranus, a star in deep space, or even an asteroid or comet instead of the moon of Venus because none of the aforementioned objects were located at any of those coordinates when the Cytherian moon was viewed. A unique object was being observed in the skies above Venus.

Searching for the moon of Venus was not then a fool's errand, nor should it now be evaluated as such. This is because such a moon most likely did once exist. It is difficult to understand how so many so-called modern experts could not develop an obviously logical reason to explain the disappearance of this moon from the skies of Venus.

The demise of the missing moon of Venus was most likely brought about when it ventured past the Roche limit, the distance from the primary planet below which a moon cannot trespass without plummeting to destruction below.

The erstwhile Moon of Venus was lost to view because it fell out of orbit and crashed into the planet.

The vanished moon was only visible to the scientists during this select time period because before then telescopes simply were not powerful enough to see it. No one saw the little moon after the early 1800's because it had fallen to its destruction by that time.

As mysterious as the moon of Venus is, so is the author of this book, Dr. Ferdinand Schorr. His name appears nowhere else in the annals of science other than as being the author of this book. My suspicion is that it is a pseudonym for one of the prominent astronomers of the day. Professor Schorr by any name certainly possessed the knowledge of an expert in this field.

The book that is to follow has been abridged. The original author spent a great deal of time describing telescopes and their locations and this is information of little use today because most of the mechanisms and locations listed are no longer in existence.

Winter Haven, August 25, 2025 G. R. Stempien

FOREWORD

The two forthcoming transits of Venus across the solar disk, which will be observed in the course of this century, will not only provide the most accurate observations for determining the solar parallax or the distance of the sun, but they will also partly decide the still unanswered question of whether Venus has a moon or not. In several astronomical writings, this discussion has been considered as already done, and the earlier appearances of this moon have been regarded as representing only negative results.

However, special investigations and comprehensive presentations of this important subject have not been undertaken up to now, since one has not followed the opinions of P. Hell or the famous author of the Cosmos, Alexander von Humboldt, who regarded those earlier observations as illusions or which they described as astronomical myths of an uncritical time, and which they attached too much importance to.

Some discoveries in the field of astronomy very often occurred amid contradictions and erroneous explanations of the phenomena related to them, for it was not infrequently difficult to grasp the difficult points correctly, and one was not always able to proceed from the existing facts to the causes. For the most part, it was also the lack of an overview of true principles, the prejudice of habit and error, through which fallacies deceived our senses.

But tireless perseverance, in its irresistible course, nevertheless leads, albeit sometimes belatedly, to the correct explanation of observations labeled as illusions or erroneous. It then demonstrates the agreement of the theory with the observed facts and leads to the ultimate goal of all research, all investigations—to the truth.

It would not be difficult to cite several examples for what has just been reported, but most of them are well known, so I may only be permitted to point out a few observations for which the correct explanation could only be found after the passage of some time.

The actual more detailed investigation of the moon of Venus was prompted by the excellent systematic compilation of all the phenomena of Venus in Dr. Klein's "Handbook of General Celestial Description."

Soon after the invention of the telescope, Galileo observed Saturn, which would later become one of the most remarkable phenomena in the starry sky, since it is surrounded by a ring, contradicting every analogy of the planets known to date. The great explorer was the first of all astronomers to see this extraordinary planet through his telescope and discovered the incredible fact that it was composed of three bodies.

However, when he later saw Saturn completely round, he discontinued his observations. Thirty years later, Gassendi published that, according to his observations, the same planet had two round bodies next to it, which often appeared elongated, but occasionally separated from the sphere and became completely invisible. There had to be an explanation for this phenomenon, but the astronomers of the time tried in vain to find sufficient reasons, especially since other observers had also discovered the same thing.

The tireless Hevelius, who made every new phenomenon in the sky the subject of investigation, also observed the strange planet around the middle of the seventeenth century and finally concluded that six different forms could be assumed for it, without adding any further explanations. Scheiner and Fontana, the leading observers of their time, also made a series of painstaking observations and each discovered special forms, such as the fifth moon.

A sheet of Doppelmaier's celestial maps, published in Nuremberg in 1742, depicts them. Finally, the astute Huyghens succeeded in proving for the first time that Saturn was surrounded by a free-floating ring, through whose various positions in relation to the Earth and its orbit around the Sun those strange phenomena on the planet were observed.

The discovery of binary stars, which currently constitute a major branch of observations of the fixed star sky and offer researchers an inexhaustible treasure trove, was initially doubted. About a hundred years ago, Christian Maier, then director of the Mannheim Observatory, published his observations of fixed star satellites, dark bodies that orbit the fixed stars and receive their light from them. The well-known Father Hell, already mentioned, enjoyed the reputation of one of the greatest astronomers at that time; he completely denied the observation of such celestial bodies, all the more so since the discoverer stated that the distance of the so-called fixed star satellites from their main star, as he observed, was within six degrees. The then permanent secretary of the St. Petersburg Academy of Sciences, N. Fuss, demonstrated in a thorough treatise that several points of this supposed discovery were based on errors, which he particularly emphasized. The reasons he put forward against it were so convincing that it should be obvious to everyone

One cannot speak of fixed star satellites in the true sense of the word here. Yet, essentially, apart from the great apparent distance between the main star and its satellite, Maier had not made any erroneous observations, for these were subsequently brilliantly confirmed by W. Herschel, who conducted investigations into this matter using the largest reflecting telescopes, and through him, astronomy received an invaluable series of the most beautiful discoveries in this field. It is well known how J. Herschel, South, the two Struves, father and son, Mädler, and others have worked on and expanded this scientific subject in the present century.

Fontana, mentioned earlier in the report on the various forms of Saturn's rings, missed no opportunity to make new discoveries in the starry sky. He was undoubtedly a very attentive and precise observer, who first discovered the spots on the planets Venus, Mars, and Jupiter. Perhaps it was not only the excellent, albeit very long, Campani telescopes that served him as an aid, but also the clarity of the sky in southern Italy that was extremely favorable. Thus, he also discovered a small, faintly luminous star near Venus, which he immediately considered to be a satellite of this planet, although no one else in his lifetime reported this new discovery.

About thirty years later, the same moon was seen again by the famous Dominic Cassini, and until 1764, it was also observed by several well-known astronomers. Some time before this last observation, the aforementioned P. Hell declared all observations of this Venusian satellite to be illusions, since they were merely reflections from the eyepiece, caused by the strong light of Venus. Lalande and the famous P. Boscowich agreed with this opinion, since they, along with several other astronomers, had searched for this satellite in vain. Later attempts to see this satellite in question during the last transit of Venus across the Sun were also unsuccessful; therefore, the lateral reflection was considered the correct explanation for those observations. This also prompted Alexander von Humboldt in his ***Kosmos*** to make a remark on this subject, which is by no means appropriate for a time when many great discoveries enriched astronomy.

Since this remarkable object has been relegated to the realm of illusions by a superficial explanation, without further investigation, I thought I would examine the observations of that time as far as the positions of Venus and other events in the sky that influenced these astronomical observations allowed. Likewise, the phenomena of the same planet were also examined in connection with the moon in question, for which there are still no adequate explanations.

This investigation was already completed several years ago, but I believe it should not be postponed any longer in order to make it known before the beginning of the transit of Venus. The subsequent transit of 1882 will also provide an opportunity to arrange for a precise observation of all the small spots on the solar disk during that transit. After this investigation, reported here in detail, I consider myself justified in concluding that the earlier observations of the moon of Venus were not illusions and that they belong to the motion of a celestial body that orbits Venus in its orbit, but is subject to such changes in its visibility that it sometimes remains invisible to us, but at other times shines with a fairly strong light. The moon of Venus belongs to the citizens of our solar system; more recent and more precise observations than the earlier ones will one day put its existence beyond doubt and provide the means to determine its orbit as precisely as the current state of science requires.

Königsberg.
The author.

INTRODUCTION.

It is not known with certainty when and where the first observations of the heavens were made, but it was only a contemplation of the immeasurable host of luminous celestial bodies that occupied the peoples of that time during the nightly hours. How often may they have turned their gaze to this or that star, whose gentle light shone down upon them, yet at that time they saw nothing of what the heavens tell the Bulim of Him who made them!

When we, contemplating the strange earthly striving toward those far-off realms, look up into the realm of order, wisdom, and incomparable beauty, we often find a consolation that reminds us of the transience of all earthly things and the eternal constancy of all heavenly things. Therefore, there is no more beautiful, no more sublime occupation for the human mind than that of astronomy, a claim so often expressed by the ancients. One of our most witty writers says: "If we were to take anything into the afterlife, it would be, in addition to our refined being, also the knowledge of the wonders of heaven."

The division of the stars into constellations was undoubtedly the first beginning of astronomical activity, for this also determined the various positions in the sky that the sun occupies in each month. The groupings of the various constellations in the firmament were sought to be linked to symbols of events on Earth at certain times. This is how the zodiac came about.

The apparent movement of the sun and moon within the same space soon led to a knowledge of chronology. Yet what a long period separates those beginnings of observations and the knowledge they yielded of the number of planets visible to the naked eye and the period of their orbits, until the appearance of Hypparchus of Samos, who elevated astronomy to a science? He first undertook to record the number of stars and to determine the distances of the sun and moon; his painstaking scientific work, which he left to posterity as an invaluable legacy, testifies to admirable perseverance and rare persistence.

As if he had not used his discoveries for further research, Ptolemy of Alexandria was determined to organize everything that had come before and to bring it into a system that would prevail for millennia. How strenuous and time-consuming the observations and calculations of these great men must have been at that time, since their instruments, the so-called armillae and the torquetrum, could not achieve great accuracy, and their observations with the naked eye deviated significantly from the truth.

They probably had no idea at the time that, after a long time, a telescope would be invented, either by chance or with the help of science, that would bring us closer to the expanses of the heavens and significantly enhance our vision. Many centuries had to pass after these rare astronomers until an insightful man, Nicolaus Copernicus, appeared who, as if endowed with a higher gift, such as is granted only to pure, noble souls, was granted the prophetic vision to clearly perceive the movement of the celestial bodies, which no one had previously been able to sufficiently fathom. He too had to struggle with all kinds of obstacles in his observations. Not only did his instruments not allow for precise observations, but he also found it difficult to observe the stars on the misty shores of the Curonian Lagoon. Despite all his efforts, he never saw the planet Mercury, which is why he is said to have bitterly lamented this failure on his sickbed before his death! If this great man had observed with a telescope back then, how often would he have had the opportunity to see by day the planet that he could never find in the morning or evening hours.

The invention of the telescope and its gradual perfection.The question of what our knowledge of the heavens would be like today without the invention of the telescope is a serious one, had Copernicus's system, which was even pursued by the papal throne, so quickly broken ground and gained prestige.

While it is well known that it is difficult to suppress truth once discovered, one can certainly assume that the correct system would have been universally recognized as such over the course of time; one cannot conclude from this that Ptolemy's system, venerated by age, would have retained its fame to this day. But how far would our knowledge have progressed without the invention of the telescope? Our knowledge and our view of the universe would now be deficient.

Almost simultaneously with Kepler's discovery of the true shape of planetary orbits, that marvelous invention occurred that significantly expanded the field of astronomy and granted greater precision to astronomical observations. This marvelous instrument was intended to guide the explorer's gaze into the deepest reaches of the universe and unfold before his eyes the most beautiful discoveries. How the first telescope came into being, however, is not as well known as one might wish.

THE MINOR PLANETS (other moons)

If the attentive observer has spent years scanning the heavenly vault in the silence of night and noted the changing forms of the celestial bodies, then it is reasonable to assume that he also knows their paths in advance and interprets the wonders of the stars with a bold vision. Premonitions arise from his soul about what his eye sees, surmises give rise to doubts; often enough, the desire may have stirred within him to penetrate deeper into the unknown realms of the heavens, that his vision might be strengthened by some aid, then it would be capable of making unexpected, as yet unknown discoveries.

For centuries, thinkers and researchers of all ages and spheres had observed the heavens; they had already discovered many truths that still exist today and left them as a precious legacy to their grandchildren. Yet it seemed as if the beginning of the seventeenth century would mark a new era in astronomy and usher in a new chapter in the great study of the heavens, revealing to the researcher an infinitely small part of the wonders of the universe and its laws. Much had already been prepared; only a small advance was needed to produce a tool to fulfill this oft-expressed wish.

Who can imagine Galileo's delight in 1610 when, after many years of observing the heavens, he first observed the stars he knew so well through a telescope of his own making? How great must have been his surprise and joy at the sight of the great wonders of creation, which he now saw as if with new eyes. And it was only this sense alone that the great Creator of the universe permitted the human race to enhance its power through an aid, while all other senses are denied this.

Yet the noblest and most beautiful of these was intended to allow us to behold the works of Him who brought everything into being in the most beautiful harmony and order. What the great martyr of truth, Galileo Galilei, saw then with his new telescope, he described in his work: Nuncius sidereus, magna longeque admirabilia spectacla pandens philosophia et astronomia, etc. (Florence, 1610, 8).

Galileo applied the telescope, which had been unusable in Holland, to the heavens and made the most beautiful discoveries that immortalized his name. He discovered that the moon, like the earth, had an uneven surface and taught how to measure the height of its mountains by their shadows. He saw the host of fixed stars infinitely increased, for in the constellation of Orion alone he counted 500 new stars and 36 of them in the Pleiades, where the naked eye sees only 6 or 7. He resolved the nebulous spot called the Manger into its individual stars and, for the first time, suspected that the Milky Way could be resolved in the same way with larger telescopes.

He also noticed the presence of Saturn's rings, although their shape did not yet give him any real idea. It is even claimed, and proven, that he saw sunspots and considered them to be atmospheric. From their common advance from east to west, he concluded that the sun's body rotated at a certain time and that its axis was inclined relative to the plane of the Earth's orbit.

This was the beginning of a great and promising era brought about by the telescope, which, through its gradual improvement and the emergence of the most ingenious men who used it, led to the most magnificent discoveries.

The World of the Jupiter

More important than all the wonderful discoveries that Galileo made in the starry sky, he considered that of the moons, which circled around the largest planet in our solar system, since they represented the Copernican world system in miniature, and they convinced him of its correctness. He wanted to name these moons Medici stars in honor of his protector, Grand Duke Cosmo II of the House of Medici, and he discusses this in his aforementioned work. The rare astronomer describes the story of their discovery there in as much detail as he was accustomed to expressing himself.

"On January 7th at 1 a.m.," he says, "as I was observing the stars through the telescope, I saw three small stars on Jupiter, which, although very small, appeared exceptionally bright. Although I initially took them for planets, I was nevertheless very surprised to see them in a straight line, and this line seemed to be parallel to the ecliptic. On the eastern side of Jupiter were two stars, and one on the western side. When I continued my observations on the 8th, I found the constellation had changed, for all three stars were west of Jupiter and closer to each other than the previous night.

On the 10th, I saw only two stars on the eastern side. On the 11th, I also saw only two, but the one nearest to Jupiter seemed to have moved a third further from its principal planet than yesterday. I believed at that time that three stars revolve around Jupiter, as Venus and Mercury revolve around the sun.

On the 12th at 1 o'clock in the morning I saw these two little stars again, but one was east and one was west of Jupiter. The little eastern star seemed brighter than the western one, but both were very clear, both at the same distance from Jupiter. A third little star also began to appear at 3 o'clock, it came from the eastern side of the planet and was very small. All these little points of light were always in a straight line, which ran in the direction of the ecliptic." The last-mentioned little star was the first observation ever made of the satellite's exit or entrance.

Since these things are so, when, under your patronage, Most Serene Cosmos, I had explored the stars unknown to all the previous astronomers, I decided with the best right to mark them with the name of your Augustus Prosapia. Since I was the first to investigate them, who can rightfully rebuke me if I also put the same name on them and call them the Medicean Stars? Dedicatory Epistle of the Star Medicii.

"On the 13th, I first saw four small stars: three western and one eastern. They formed almost a straight line, for the middle of the three western ones was slightly to the north. The small eastern star was two minutes from Jupiter, the others were only one minute apart. They were all almost the same size, but brighter than small fixed stars. The 14th was dim. On the 15th, at 3 a.m., I drew, as usual, the constellation of these small stars. There were four of them, all western.

Their position was in a straight line extending somewhat to the north." Galileo published his observations up to March 2 of that year and thus became convinced that Jupiter had four moons that orbited the sun from west to east, like the planets around the sun. On March 28, 1611, he informed a friend that he had already discovered the orbital periods of the two outermost satellites and had calculated tables based on them so that one could know at any time what distances these two satellites were to the right or left of Jupiter.

Perhaps, he said in his letter, I will soon discover the orbital periods of the other two as well. Galileo is known to have continued these observations until he became blind, determining the orbital periods of the moons. However, better and more accurate are the determinations of Dom. Cassini, who calculated the first usable tables that perfectly corresponded to the sky. Galileo observed the eclipses of these satellites, or their entry and exit from Jupiter's shadow, as well as their transits from east to west across the planet's disk. He also noticed how they sometimes suddenly disappear on the western side of the planet, even before they were occulted by Jupiter, which represented their entry into its shadow:

Simon Marius claimed, as already mentioned, to have seen these moons as early as the end of November 1609, which may be probable according to what has been reported regarding the invention of the telescope, but Galileo, even if he himself were not the discoverer, deserves the credit of having first observed them astronomically. He also published his long-continued observations of these satellites and determined their orbital periods, which are identical to those of Cassini's later calculations, were in good agreement. Marius, however, also calculated tables of these moons, which soon became unusable because they deviated significantly from the observations after a while. Since the position of the four moons forms a very slight inclination with the ecliptic, or Earth's orbit, the first three are eclipsed with each orbit, while the fourth often passes across the disk of Jupiter.

The transits of the satellites have been observed very carefully, for, during this time, the Moon casts its own cone of shadow on the surface of Jupiter, obscuring a small portion of the disk illuminated by sunlight, on which the shadow is perceived from Earth as a round spot. However, a second, equally large gray spot is seen on the disk, preceding the one mentioned above if this transit occurs before Jupiter's opposition, or following it if this phenomenon occurs after the opposition.

With very good, high-power telescopes, such a moon is sometimes seen as a dark, no longer round, but irregularly defined gray spot, considerably smaller than that of its shadow. Harding and Schröter concluded from this that these moons must sometimes have very large and dark patches on their surface or in their atmosphere; however, the strong illumination of Jupiter, which casts its light on one of these moons, can also exert its influence in this way, causing the contours to become indistinct. because it is assumed that this light is reflected more from some places on the orbit of the small moon and less from others.

In 1797, Herschel presented a paper to the Royal Society in London containing the results of his numerous observations on the intensity of light and the relative size of

Jupiter's satellites. The main result of this investigation was that the aforementioned intensity of the satellites is highly variable, and that their apparent size is not always constant. This led the famous astronomer to the conviction that the variable sizes, as well as the non-constant intensity of their light, must have their origin in the physical nature of these celestial bodies. He demonstrated in a surprising way that these moons are covered with spots on their surface that reflect light to a greater or lesser extent, and that they move around themselves.

It was, however, of great scientific interest to determine those places in their orbits where such a great change in light occurred, and Herschel used a graphical operation for this purpose, which soon convinced him of the regularly recurring phenomena of these satellites. He noted on the four circles representing their orbits the places where, over a long period, each satellite had appeared at its greatest or least luminosity.

He also marked the places where each satellite had appeared at its greatest or smallest apparent diameter, from which he subsequently discovered that these phenomena always recurred in the same regions. The astronomer further concluded from this that these moons possess a rotation around themselves that is equal to the period of their orbit, as is the case with our own moon.

The first satellite shows this change to the greatest degree; it is at its greatest brightness when it reaches the point almost halfway between its eastern deviation and conjunction. The second satellite's most brilliant phase is towards the Earth when this small star is between its greatest eastern deviation and conjunction. The third satellite has two periods of brightness, and it can be observed at both deviations. The fourth satellite's light remains essentially the same everywhere and is quite dim, but after opposition its light becomes somewhat brighter. Herschel published a detailed treatise on this under the title: "Observations of the Changeable Brightness of the Satellites of Jupiter, and of the Variation in their apparent magnitudes; with a determination of the time of their rotatory motions on their axes etc. London 1796.

Furthermore, the famous Slough astronomer found that the color of the first satellite is more or less vivid white, but Beer and Mädler in 1836 found it more bluish; according to other observers, it is yellowish, thus appearing to change color. In contrast, the second appears sometimes pure white, sometimes greenish white, and often bluish.

Flaugergues, who observed this moon quite often, considers it a minor planet, which changes its light most rapidly. The third satellite is always white, while the fourth sometimes appears very dark, or orange, or reddish. It is not easy to adequately understand this strange color change.

The question has often been raised as to whether it is possible to observe these moons with the naked eye; they would be visible even if the bright brilliance of Jupiter did not outshine them and render them invisible. They disappear to the eye, yet a telescope of insignificant magnification reveals them as small stars. Arago therefore conducted experiments with telescopes without magnification, which in fact represent a flawless eye. These experiments demonstrated the possibility of observing these moons without the aid of any optical instrument. Using this experiment, he and all the astronomers at the Paris Observatory immediately saw the third moon not far from the planet.

Dr. Klein, in his well-known work, "Handbook of General Astronomy," page 162, however, denies such visibility and says very aptly: "I believe that in cases where a satellite was seen with the naked eye, two satellites were actually in close conjunction, and the intensified impression of both was sufficient for visibility. Professor Heis in Munich has proven through numerous experiments that his eye is one of the sharpest in existence. The stars appear to him completely without false rays, which is not the case for an ordinary eye, even when, like Arago, it uses a small, non-magnifying telescope.

Nevertheless, Heis has never succeeded in seeing a satellite with the naked eye. On April 11, 1863, at 1:15 p.m., in favorable weather, he saw a satellite clearly to the right of Jupiter, and between it and the planet a noticeable gap. However, investigations showed that the perceived point of light was the union of Jupiter's third and fourth moons, both of which were at a distance from the main planet equal to seven times Jupiter's diameter.

Already some earlier observers, such as Pound in 1719 and Messier in 1767, observed the first and second moons appearing brighter and more luminous than the limbs of the main planet during their entry. This was also recently noticed by Chacornac through careful observations, and these observations were confirmed. Maraldi claimed to have observed the change in brightness of the third moon as early as 1704, even claiming to have seen spots on its disk. Messier, one of the best observers of his time, also published a similar observation in 1768; however, the spots were only visible when the moon was in front of Jupiter and its edges could not be seen on its disk.

This was fully confirmed by repeated observations by the same astronomer on September 6, 1760, and July 14, 1771, and the presence of the spots was confirmed. Recently, Secchi was able to observe the spots again and came to the conclusion that Jupiter's third moon is an exception to the others in terms of its rotation period, which is significantly shorter than its orbital period.

The fourth moon often appears to be the smallest or faintest, as the observations already cited by W. Herschel demonstrate, yet at times it equals the third moon in the intensity of its light, and at other times it surpasses all other moons in its brilliance. Incidentally, the elder Herschel's observation was also confirmed by Beer and Mädler, for they always observed any changes in light after the course of a synodic orbit.

The World of Saturn

Galileo's telescope had neither sufficient magnification to observe several now-known objects in the starry sky, nor was its construction suitable for clearly visualizing Saturn's ring and confirming its position around the planet. It was therefore impossible to discover Saturn's satellites through it.

Since they are only visible to the observer through good optical instruments when Huyghens, the famous Dutch astronomer, observed Saturn on March 25, 1655, with one telescope of 12 feet and another of 23 feet focal length, he saw for the first time its fourth moon. This is the largest of all the planet's companions and the only one visible in an ordinary dioptric telescope of 10 to 12 feet.

Dominic Cassini discovered the fifth moon towards the end of October 1671 with such an instrument of 17 feet; the third moon appeared only with the use of a stronger telescope of 35 or 70 feet on December 23, 1672, about which he published a short treatise.

In March 1684, he observed the two inner moons, that is, the first and second. This discovery was made with Campani telescopes of 34, 47, 100, and 136 feet. He also confirmed it with Borelli telescopes of 40 and 70 feet, and with Artonquell's, which were even longer. In England, the existence of these four moons, which Dom. Cassini had observed, and was doubted, but the astronomer Pound, known for his precise observations, sought them out. For this purpose, in 1718, he had the excellent objective with a focal length of 123 feet, which Huyghens had sent to the Royal Society in London, installed above the bell tower of his parish. He observed all five moons, and based on these observations, the elements of their orbits, which Cassini had calculated in Paris in 1714, were improved. At the same time, when the then Vice-President of the Royal Society Society, Hadley, had found a means of manufacturing good reflecting telescopes according to Newton's principle, they continued to use them to observe Saturn's satellites.

The first and second of these are difficult to discern with ordinary telescopes of 40 feet focal length; the third is somewhat larger, and is sometimes seen throughout its entire orbit. The fourth is the largest of all, which is why it was discovered first. The fifth surpasses the first three in luminosity when in its eastern ebb, but is sometimes very small and disappears entirely. As soon as achromatic (lenses not distinguishing colors - trans.) telescopes were invented, their observation became easier, and the often-mentioned long telescopes were no longer needed.

The well-known Swedish astronomer Wargentin in Stockholm assured that he had seen them all through such a 10-foot telescope. After Huyghens's first discovery of a satellite of Saturn, it was rightly suspected that there might be more; it was believed their number must be equal to the number of planets in the solar system.

But the Dutch astronomer did not seek them out; perhaps he did not consider his otherwise excellent telescope, whose lenses he had ground himself, sharp enough for such observations. Later, with the help of his objective, he discovered Po and all the other satellites known at that time. However, it is more likely that his other great discoveries and inventions prevented him from making these sustained observations. For a long time, these five moons seemed to form the complete system of Saturn's minor planets, until on August 28, 1789, Herschel, with his large 40-foot telescope, observed one that was even closer to the ring of the main planet than the fifth. This small star should therefore have formed the first in the series of these moons, and they would now be one unit further away from Saturn in their nomenclature. To avoid confusion in naming them, the previous designation was retained and this one was called the sixth moon. Likewise, through that giant telescope, the tireless observer succeeded in discovering the seventh moon on September 17, 1789, located between the ring and the aforementioned sixth companion.

This last-discovered moon is extremely faint, and Herschel could only discover and observe it under favorable circumstances; in such cases, it was visible through the 20-foot telescope. After Herschel's discovery, the number of these moons was considered known, so it was all the more surprising when, on September 16, 1848, Bond in America discovered an eighth companion, which is probably the last in the moon system of that rare planet, if discoveries are listed chronologically. (There are now at least 30 - Trans.).

The orbits of the seventh and eighth satellites are very small, often passing behind the main planet or passing in front of Saturn's disk. Sometimes the light from Saturn's ring makes them completely invisible; if the ring appears to us as a luminous line that is incredibly narrow, these moons are observed along it as small, moving, shining dots. A rare exception to all known orbital periods in the solar system is that of the seventh satellite, which is less than a day, but the firmament offers so many wonders to the human eye that this is not the greatest.

The observations that Cassini made with his large telescopes in 1684, and thus discovered the two moons, were made with two lenses from Campanile, one of which had a focal length of 100 and the other 136 feet. He subsequently also used two new objectives of 90 and 70 feet. These large glass lenses were used for observations without an I-tube, as previously mentioned; he placed them sometimes on top of the Paris Observatory, sometimes on the top of a large mast, and finally on a wooden tower that had been brought from Marly to Paris for this purpose.

Herschel also spent some time trying to determine the luminous intensity of these satellites, although he encountered several difficulties here than with those of Jupiter. He noticed that the fifth moon is at its brightest when, after passing through inferior conjunction, it is at the 68th and 129th degrees of its orbit—assuming that the latter is counted from the conjunction—in this interval it is almost less luminous than the fourth moon. Earlier, in 1705,

Cassini and Maraldi observed that the satellite disappeared completely when it was to the east of its celestial body. The variation of this fifth moon can be compared very well to that of a star that, to the naked eye, decreases from the second to the fifth magnitude, as we often observe with variable fixed stars.

William Herschel observed these gradations of brightness during more than 10 orbits and concluded that this moon always presents the same side to its principal planet. Some observers contemporary with Cassini expressed the opinion that in September 1705, the fifth moon was seen equally east and west of the principal planet, and it was concluded from this that its rotation period around itself was equal to its orbital period around Saturn. This conclusion would not be valid if the moon had gradually been at its brightest in different parts of its orbit. Its reappearance in the east on rare occasions is believed to be the sign of a physical change.

The shadows of Jupiter's transitory moons were often seen on its disk; Herschel made an analogous observation with regard to Saturn's moons when he saw very clearly the shadow of the fourth moon on Saturn's disk on November 2, 1789.

To avoid any misunderstanding when mentioning these moons, they have been given the following names:

Mimas.

Enceladus.

Thetis.

Dione.

Rhea.

Titan.

Hyperion.

Japetus.

Mimas was also later observed by Schröter; de Vico in Rome saw it more recently on June 27, 1838, from which, in conjunction with Herschel's observation, he accurately calculated its orbital period. This moon is difficult to observe, but according to de Vico's observation, it is better seen if Saturn and its rings are covered by a device attached to the objective. During the disappearance of the ring in 1802, all eight of the planet's moons were seen at Pulkovo.

Since its discovery, Lamont in Bogenhausen also observed Enceladus during the summer months of 1830 and thus determined its orbital period more accurately.

Thetis was seen again by Herschel and Lamont at the end of the seventeenth century, and the latter astronomer calculated from these observations its orbital elements.

Dione was also seen by Herschel the Elder and later by his son, John Herschel, during his stay at the Cape of Good Hope from 1835 to 1837.

Ithea was seen by Schröter and later by John Herschel.

Titan, the first moon discovered, was observed for some time by Mädler (though discovered by Huyghens - Trans.) among others, and its diameter was determined by micrometric measurements, from which it was calculated to be 350 to 400 miles. Its orbit was determined by Bessel more accurately than that of any of the planet's other moons.

Hyperion is even fainter than Mirnas, but recent observations made in Malta do not seem to confirm this opinion of several astronomers.

Japetus was again observed by Herschel, the Elder and the Younger, but its orbit has not yet been calculated very completely, since more observations of this moon are still pending.

The World of Uranus.

The immense distance of this planet, its small diameter, and its weak luminous intensity gave little reason to hope that its moons, if it possessed them, could ever be observed. Considering the relative size of Jupiter's moons to that of their parent planet, or that of Saturn to this celestial body, since these two are the largest planets in the solar system, similar companions of Uranus should not be visible to any observer. But the discoverer of this planet was not the man to be discouraged by such conclusions. Powerful telescopes of ordinary construction, or telescopes based on the Newtonian principle with two combined mirrors, would not have led him to any discovery; he replaced them with his front-view telescope, described by us in January 1787, namely, with a reflector capable of transmitting more light to the images of the observed objects, because the small reflector was not included in the telescope, thus eliminating any diminution of light.

On January 11, 1787, when Herschel used such a telescope, the dimensions of which were not known but were probably those of a 20-foot telescope, he saw Uranus surrounded by some very small stars. Their position relative to the planet was noted with all possible care. On repeated observation the following day, two of these small stars had disappeared. This sign of the probable existence of Uranus's moons prompted the very precise observations of January 14, 17, 18, and 24, as well as those of February 5, and especially February 7 of the same year. On this last day, the famous astronomer kept his eye on the telescope for nine hours without a break and had the satisfaction of observing a satellite gradually moving along its orbit. The presence of a second moon was not confirmed until the third day later, the 9th. Although the series of observations continued until February 11, it was still not possible to determine with great precision the time of the revolution of these two moons in their orbits.

However, Herschel was able to conclude from them that the synodic revolution of the first satellite, the one closest to Uranus, was completed in 8 3/4 days, while that of the second was estimated at almost 13 1/2 days. He added to these approximate determinations the observation that the planes of their orbits formed significant angles with the ecliptic.

The famous observer tirelessly searched for new companions of Uranus, although their observation was very difficult. In a treatise published several years after these reported observations, namely on December 14, 1797, he reported the discovery of four new moons of the same planet, bringing the total number to six. The relative positions of these moons and the date of their discovery were as follows:

1. Moon, the closest to Uranus, discovered Jan. 18, 1790.

2. The closer of the two previously discovered 11 Jan. 1787.

3. May 26, 1794.

4. The more distant of the two previously discovered 11 Jan. 1787.

5. Fbr. 9, 1790

6. The most distant of all Fbr. 28, 1794.

Only an astronomer like Herschel the Elder could overcome so many difficulties, not only in astronomical observation, but also in the perception of almost invisible small stars. Often, he mistook very small fixed stars, which happened to be in the vicinity of a planet, for its moons until he realized his error. Therefore, in his second treatise on this subject, he did not dare to settle the question of the duration of periodic orbits. However, to somewhat satisfy the curiosity of astronomers, he published the following results:
Duration of the orbital times:

1st month 5 days. 21 St. 25 minutes.

107 days 16 hours 40 minutes at least.

Derived by Kepler's third law from the already determined orbital period of the fourth and the assumption of the greatest distance in the orbit of 25.5". Determined in 1787.

Derived by the hypothesis that this moon is located exactly in the space between the second and fourth in the middle, or that its distance from the center of the planet is 38.6". refers to what was predicted.

Derived from an observation of the distance of the moon from the center of its planet, from which the latter was seen to be five times farther than the fourth.

According to an observation, this sixth moon is at least four times more distant than the fourth.

The results reported here do not inspire much confidence, and yet they have the appearance of an accuracy that can only be achieved through effort, for one rarely achieves such precision in calculation that one can give a number down to minutes, when this requires quantities that still contain errors of several hours.

What is important in the aforementioned treatise of 1797 consists in the sentence, which incidentally was transmitted without reservation: "I take this opportunity to indicate that the motion of the satellites of George (Uranus) is retrograde." W. Herschel's third and last treatise on the moons of Uranus is dated June 8, 1815. Several astronomers seriously doubted the existence of the four moons indicated in the treatise of 1797.

With the help of various observations made during the year 1798 that were undertaken, Herschel added new arguments to those he had already published; these relate to the intrinsic motion of the same four stars which he discovered around Uranus between 1790 and 1794. It is, however, very clear that his intention was to keep within the bounds of probability in this respect.

Regarding the two moons first discovered in 1787, Herschel perfected his theory in his 1815 treatise, and only through painstaking investigation did he arrive at the discovery of their synodic orbital periods, which he had previously only reported in fractions of a day.

He marked the ascending node of the orbits at 165° 50', while their inclination to the ecliptic was 78° 58'. Finally, he reiterated his statement of 1797 regarding the direction of motion of these moons. Without ambiguity, he published a result that is particularly noticeable to us as the only anomaly for motion in the solar system, if we exclude comets: "The two moons of Uranus move around it in retrograde motion, or, what is the same, these moons describe the northern arc of their orbit, which lies between the ascending and descending nodes, by a motion directed from east to west.

In Herschel's opinion, the first moon, the one closest to the planet, was undoubtedly more luminous than the other, a fact he occasionally verified. However, he later changed his mind. By constantly observing these two moons, the tireless explorer believed he had discovered that their relative luminosities varied considerably, and that the second moon often surpassed the first in this respect.

These variations, according to Herschel's opinion, are caused by their rotation, from which the result is that during each orbit around the planet, these small stars gradually reveal to us the various parts of their surface, which are undoubtedly unequally illuminated. But if these variations in luminous intensity were not regular, they are caused by the atmosphere, which gradually covers or exposes some more or less large parts of the bodies that are dark below.

From the totality of the observations contained in the 1797 treatise, one can conclude that these moons are never visible if their angular distances from the planet are below certain numbers. For the first moon, this limit in distance is 14", whereas for the second, it is 17".

Similar changes cannot be attributed to the influence of Uranus's atmosphere, because the phenomena occur equally whether the moon is on the part of its orbit closest to us or further away. Nor would there be any reason to seek the cause in the movement of rotation, for these observations show that at certain distances from the planet, the moons exhibit phases that are relatively dark. Assuming all circumstances to be equally effective, it follows that even the small fixed stars disappear at the same distances from the planet as the moons.

The confirmation of the optical rule is evident here, as everywhere: "A great light prevents those faint lights that are near it from being seen." However, in this case, the peculiarity of Uranus appearing as a great light must be considered.

The question of the cause of the luminous glow of Uranus, which makes the moons and the minor fixed stars invisible, could still be raised. Is it located at the focus of the telescope as a result of the aberration of the mirror, or does it appear in the eye due to the absence of glare on the cornea? To resolve this question, it would have been sufficient to cover the planet with a thin thread at the focus of the reflector; perhaps in this way, the moons might have been seen at distances where they previously disappeared when the planet's disk was visible. It is regrettable that W.

Herschel did not concern himself with observations of this kind.

Herschel's treatise would have been incomplete if he had not continued it by communicating the measured greatest deflections of these two moons. However, nothing seemed more suitable than these observations to furnish proof of how many difficulties these truly microscopic stars presented to the astronomer, for having reached this point, he himself seems to have been content with the communication to hesitate; the lack of agreement in his own results was unpleasant to him. However, forced to make some decision, he remains with the epoch of his observations and reports, according to his investigations, as actually occurring deviations of 36" for the first moon and 48" for the second.

Only a short addendum would suffice to summarize what we have learned about the moons of Uranus since the aforementioned work of Herschel. The discoverer's son, John Herschel, using his own observations from 1828, 1830, 1831, and 1832, found the following periodic orbital periods of the two moons of 1787, the only ones that could be observed with a 20-cracked reflector whose mirror had been repolished.

These investigations date from 1834 and appeared a year later in the eighth volume of the collections published by the Astronomical Society of London. Another more recent treatise on this subject appeared in the Transact. Phil. of 1838. It was written by Lainont, the director of the Royal Observatory at Bogenhausen, stating that the German astronomer believes he has noticed that, according to his observations, the two excellent moons exhibit distinct traces of an elliptical orbit, but he has reserved the more delicate investigations on this subject for another time. His present calculations, as well as those of the two Herschels, were based on a circular hypothesis. Lamont's treatise contains a statement which is not communicated in much detail, but which nevertheless, in relation to the outstanding merits of this astronomer.

He must have attracted the attention of observers, for he says that on the evening of October 1, 1837, he saw and observed the sixth moon of Uranus. Thus, it seemed that one of the four moons announced by Herschel in 1797, and which were considered doubtful, was placed in his light.

The mass of Uranus, which Lamont deduced from his observations of the two main moons, is 21600 V; this is one-quarter smaller than the value calculated by Bouvard in Paris from the perturbations produced by Uranus. For these observations of the moons, the two Herschels used telescopes of 20 and 40 English feet, while Lamont made them with an achromatic telescope of 15 feet focal length and 10 1/2 inch aperture, constructed in Munich.

John Herschel gave astronomers a criterion by which they could decide whether their instruments had the appropriate magnification to observe these moons, so that they could search for them with some success.

Between the stars ß 1 and ß' 2 of Capricorn, namely, toward the middle of their interval in a straight ascension, a little to the north of the line joining them, there is a double star composed of two stars of the sixteenth and seventeenth magnitudes, both 3 inches apart. Any instrument that does not clearly show these two stars of the sixteenth and seventeenth magnitudes cannot be used to observe the moons of Uranus. John Herschel, in fact, found this explicit remark in his observation journal, without indicating the date: "The two conjoining stars of the double star situated between ß 1 and ß 2 in Capricorn are brilliant objects compared with the satellites of Uranus."

The same astronomer further claims that this observation requires a magnification of at least 300 times, no matter how large the aperture of the telescope and its power to penetrate space may be.

The elder Herschel was not so decisive on this point, for it seemed to him that a magnification of 800 times and more was not absolutely necessary for the constant and continuous observation of these moons. He discovered them with a magnification of 157 times, but he only saw them for moments, like sparks. Their continuous observation was only possible for the discoverer with magnifications of 300, 600, and 800 times.

Lassel in Liverpool, whose excellent reflecting telescopes were already mentioned in the previous section, discovered two new moons of Uranus in October 1851, which would justify assuming eight companions of that planet. Recently, the same astronomer has resumed observations of Uranus's moons in Malta, greatly modifying the elder Lassel's discoveries, for he finds only four satellites and believes that Heracles mistook small fixed stars for the other satellites. The famous Slough astronomer has never seen the two moons observed by Lassel, as they are close to the main planet, even closer than the first moon discovered by him on January 18, 1790.

Lassel claims to have proven the existence of his newly discovered companions through his observations and believes that no more than four moons orbit Uranus. Since he has often seen them by moonlight, he asserts that no others can exist. Whether Lassel is absolutely right here will be decided only by the future, but one must trust his observations and calculations. The names given to the four moons, as they follow one another in their distances, are:

1. The moon discovered by Lassel in October 1851 is called **Ariel**.

2. The moon discovered by Lassel in Oct. 1851 is called **Umbriel**.

3. The moon discovered by W. Herschel on January 11, 1787 is called **Titania**.

4. The moon discovered by W. Herschel on January 11, 1787 is called **Oberon**.

The mean distances of these moons from the center of Uranus are: 1479", 20-61", 33'88", 45-20", from which Hind calculated for the mass of the main planet V 20,600, a value that does not differ significantly from that of the early Lam 011 t.

The World of Neptune

The difficulties in locating the companions of this planet undoubtedly seemed considerably greater than those of Uranus, but Lassel and Challis discovered a moon of Neptune in the early days of August 1847. This important discovery was made using a 20-foot reflecting telescope with a magnification of 205 to 300 times, and was later confirmed by Bond's observation on September 16 of the same year.

However, even earlier, the same companion was discovered in Pulkovo, on September 11, and observations continued until December 20, 1847, giving an orbital period of 5 days, 21 hours, 4 minutes, 9 seconds, and an average distance of 17.95 seconds from the center of the main planet.

Bond also made a series of observations of this moon, which yielded 5 days for the aforementioned dimensions. 21 hours, 5 minutes, 50-6 seconds, and 1G'3"; thus only slightly different from Struve's data.

Three years later, on August 14, 1850, Lassel, using a magnification of 628 times, saw a second moon of Neptune, but he did not find it in his later observations in Malta, which is why he also emphatically denied the existence of this satellite, although Mitchell had also seen one.

According to observations, the moon of Neptune, discovered in 1847, must be considerably larger than that of Uranus, because it is not as difficult to find as the latter, since its distance from us is considerably greater than that of the moons of Uranus.

The supposed moon of Venus.

The investigation into the existence of a moon of Venus cannot yet be considered complete, although there was a time, and it is not long past, when the idea of such a moon was decisively rejected. However, it cannot be denied that this was a hasty conclusion, which gave rise to a well-known phenomenon in the cosmos. For many years, scientists have been occupied with the question of whether Venus is accompanied by a satellite, and whether it is luminous or dark. Famous astronomers and mathematicians were convinced of its existence, while others, no less distinguished, denied it and called its observations illusory. Such disputes often have great appeal for some, for they offer a welcome opportunity to make so-called witty remarks about a science that astonishes many with its consistency and certainty.

The sublimity of astronomy is in no way diminished when some philosophers and writers hold up this moon of Venus to astronomers, as if it were a symbol of the error from which their science is not free, although it boasts of its wonderful standpoint. The discoveries in the universe, which offers us the most magnificent and diverse discoveries.

The investigations into the phenomena presented by the heavens are not as easy as one might think, for nature has no mercy for serious researchers in her field, and therefore the solution of many questions is often difficult. But if it were possible to reveal, by means of some system, most of the secrets of nature's workshop, we would undoubtedly be less encouraged to make the heavens the object of our continued observations and investigations. For all ages, the most beautiful discoveries will still present themselves to our eyes there, and there will never be a lack of opportunity to add something further to the known wonders of the universe.

The most famous astronomers of their time, who had immortalized their names in science through their magnificent discoveries, also discovered this moon. Through their excellent investigations, their precise observations, and painstaking calculations, they had already made their extraordinary achievements known to the learned world when they announced this unusual discovery. It certainly warned them to be cautious about protecting themselves from deceptions of any kind, a deception they could not avoid even after years of continuous observation. If only one famous astronomer had made known to us the discovery of a celestial body that was subsequently not seen, one might perhaps be more likely to express reservations about it, but there were several.

There must have been a special reason that could justify us in the observation that those great discoverers were ignorant of their instruments and had not become aware of a flaw that could have led them to self-deception: namely, the reflection of a bright star on the eyepiece, which could produce a secondary image due to the different position of the observer.

In some cases, this would be considered a star, specifically a new celestial body, and this is how this moon of Venus was discovered.

The first discoverer of this moon, **Franz Fontana** of Naples, was, after Galileo, one of the most distinguished celestial observers, who observed the wonders of the planets with excellent telescopes. In 1636, he first saw a spot on Mars, which he observed again two years later and was thus led to the rotation of this planet.

Even earlier, namely in 1633, he discovered the dark and light, zebra-like stripes, which were constantly, although with some variations, pointing toward Jupiter and circulate parallel to each other from east to west around its sphere. Dominic Cassini is usually considered the first discoverer of the spots on Venus, but as early as 1645 Fontana had observed them on the disk of this planet, as

Riccioli provides a picture of in his Almagast. All of Fontana's discoveries were later confirmed by Cassini, so no deception has occurred.

No one in earlier times observed as many satellites as Dominic Cassini; his precise tables of Jupiter's moons earned him great fame, which was further increased by the discovery of four of Saturn's moons, not to mention his other great services to astronomy.

Already mentioned in the first section of this work was Short, the most famous optician of his time. He was also an outstanding astronomer who participated in all important observations. This is demonstrated by his excellent micrometric measurements of Jupiter and the resulting determinations of the planet's oblateness. He observed the two transits of Venus in the last century, and he also determined the diameter of the sphere of Venus.

Montaigne discovered the comets of 1772 and 1774, as well as the second one in 1780, and made valuable astronomical observations for many years.

Christian Horrebow was director of the Copenhagen Observatory since 1753, which his father Peter Horrebow had headed until then; both are known as precise and careful observers.

Andreas Meier in Greifswalde was, as is evident from several of his astronomical investigations, a very precise and conscientious observer; in addition to his observation of the moon of Venus, we will also report another observation of Venus made by him in this work.

Rödkier in Copenhagen and Montharron in Auxerre are less known for their earlier astronomical work, but their observations presented here deserve confidence, as they agree with other contemporaneous ones.

These were the respected men who had made the observations of the moon of Venus known to date, and who, through their excellent discoveries, had already given evidence to the learned world that the information they had published

Even in our time, they would still be considered among the most deserving men of science, whose conscientious investigations are aimed at protecting themselves from self-deception and pursuing the path that alone leads to the truth. Bernoulli therefore aptly states in the Monthly Correspondence, Volume V, p. 343: "It seems almost unbelievable to me that Cassini, Short, Montaigne, Baudouin, and others could not have distinguished something such as the reflection of the eye must represent, which is actually only a shadow of a star, from a real star."

Nor can one deny the existence of this moon by citing the reason that it has not been seen since 1764. This question can be answered very simply by the fact that after the failed attempts to observe it on the solar disk during the last two transits and the position of the Venus module calculated by Lambert at the time of the conjunction of June 1, 1777, no one has bothered with such observations again. Moreover, since the physical composition of its atmosphere often renders it invisible, only a few astronomers will be able to observe it in the future, assuming that its position in the sky cannot yet be accurately calculated in advance.

The first discovery of the moon of Venus occurred on the evening of November 15, 1645, by Franz Fontana in Naples, who had already noticed a small dark spot in the center of the crescent moon a few days earlier, namely on the 11th of the same month. On that day, November 15, a dark sphere appeared just above the northern horn of Venus, and below the first, a second dark one. The discoverer could not interpret this phenomenon other than that he believed there was some aerial phenomenon, a meteor, or something similar, in front of the crescent.

After Fontana's observation became known, P. Grimaldus, Riccioli and Gassendi searched for the faintly luminous sphere mentioned above, but without discovering it in any way using various telescopes.

This passage is found in Riccioli's Almagestum novum, Volume I, p. 485. The French astronomer, Peter Gassendus, also mentions it.

A hitherto little-known observation, which surpasses the previous one in its precise timing, is the following: “Excerpt from the diary of the Greifswalde Observatory by Prof. Meier.

On May 19, 1759, at true noon, the pendulum clock showed 23 hours 55 minutes 33.1 seconds. Venus passed through the circle of noon at 2:7:45 a.m., clock time, or 2° 12' 12 1/2 " true time. On May 20, at true noon, the clock showed 3° 55' 32-5". At 8:45' 50 p.m., I saw above Venus a sphere of much less brilliance, about 1 1/2 times the diameter of Venus from it.

Future observations will show whether this sphere was an optical illusion or the satellite of Venus. The observation was made through a Gregorian telescope with a focal length of 30 inches. I continued it for half an hour, and the position of the sphere relative to Venus remained the same, although the direction of the telescope was changed.

On May 21st, at true noon, the pendulum clock showed 23 hours 55 minutes 35 seconds 4 minutes. (Astronomical Yearbook 1778, page 186.)

The following list contains all the observations of the satellite in question known to date, which, as we shall note below, deserve the confidence to base further calculations on.

Observations of the moon of Venus.

May 4, 25 or 26 Above Venus 45" As before Montaigne 9 minutes tilted towards the north.

According to observations, the moon of Venus was observed by twelve observers from its appearance until 1764, and their statements about the shape and diameter are all consistent.

Notes:

2nd observation: Cassini discovered Saturn's third moon on December 23 of the same year using telescopes from 35 to 70 feet. 3rd observation: Two years later, in March, the same astronomer discovered Saturn's first and second moons again using telescopes from 34, 47, 100, and 136 feet, thus making this scholar very experienced in the search for such small satellites.

Observation 4. Short observed this moon for a full hour using various magnifications; the satellite remained in the same position in the sky throughout.

Observation 6: From May 3 to 11, 1761, Venus advanced 2° in its apparent orbit; the Moon always appeared 20 to 25' south or north of it.

Observation 10. From March 3 to 29, 1764, Venus traveled 31° in the sky in its orbit; the moon always appeared very close to its principal planet. On March 4, at 6:26 a.m., Bödkier in Copenhagen saw the moon, which he had first seen in the evening, for the second time. An hour later, as he could no longer recognize this moon (he believed because of the northern lights that had occurred at the time), he found two stars near Venus, one of which, at the beginning of the 20th century, was considered to be Uranus by Dr. Koch in Danzig.

In fact, an unnamed mathematician had suggested in a letter to Bernoulli dated November 10, 1781, that perhaps through various encounters between Uranus and Venus, the latter had once attracted attention and, through its always brief appearances, had given rise to the appearance of a satellite of Venus.

As a result, Dr. Koch undertook an investigation of the orbit of that planet and found that on March 4, 1764, Uranus was only 10 1/2 from Venus. This distance significantly exceeds the distance of the moon of Venus at the same time, as stated above; therefore, at that time, Uranus was not considered a minor planet. The stars seen by Rödkier near Venus on March 4, 1764, were, as I was convinced by a casual investigation into the orbit of this planet at that time, two fixed stars that also appear in Meyer's Zodiacal Star Index, whose positions for 1800 were as follows: the planet Venus was approximately 7' south of them; the positions of these two fixed stars also appear in more recent star catalogs.

Given the large number of newly discovered asteroids now known, one might be led to similar assumptions as those made by the unnamed mathematician about Uranus, for one or another of the asteroids close to Venus could also have been considered such a companion.

However, this is contradicted by the impossibility that such a small planet would have remained constantly close to Venus in its apparent orbit, since in 1761 Venus advanced 2° in 8 days. Likewise, it is not likely that in 1764, when Venus traveled 31° in 26 days, an asteroid would have followed it in its apparent orbit. One can also add that all observers saw the small moon of Venus as having the same crescent-shaped form as Venus, which would not be said of the asteroids.

The first discovery of a moon of Venus, which occurred more than two hundred years ago in the clear southern sky of Italy, was viewed by Fontana's contemporaries in the same way as we now judge it, for they searched in vain for the newly discovered star. It is likely that the discoverer later also searched for this moon without finding it, for when Galileo first observed Jupiter's moons, he could observe them every night under clear skies.

Apart from the position of Venus's orbit, which prevents this planet from being observed in the meridian in the evening or at night, there must also have been some special difference in the position of this minor planet's orbit that makes it so rarely visible. A special physical structure of the surface of the supposed moon could also be the cause.

The second observation of this moon 27 years later by Cassini, as well as a similar one 14 years later, turned the attention of astronomers more to this object, but they searched for it in vain, because neither in the immediate vicinity of Venus nor far from it was any moon of the same visible, and even the famous second discoverer never succeeded in finding the same.

From these observations, Cassini concluded that Venus's newly discovered moon rarely reflects enough light to be visible.

Lambert remarks: "In fact, one sees no reason why all planets and satellites should be brilliantly bright or constantly shine with full light. After all, there are fixed stars that are only occasionally visible." The third discoverer, Short, made a very precise measurement of the distance and position angle of this moon in 1740, so his observation is highly valuable; he also observed this phenomenon for a longer period than the previously mentioned astronomers.

Until then, only isolated observations of this moon had been made, so no orbit could be determined from them. Montaigne, on the other hand, observed it four times in 1761, and his observations served to learn something more about the motion and orbit of this moon. The French scholar Baudouin read a treatise on this subject at the Academy of Sciences in Paris, which was soon published, but Berlin also did not neglect to translate it into German, appearing under the title: The New Moon of Venus.

Since the transit of Venus across the solar disk was expected in the same year, the question arose whether this moon would be observed before, during, or after that rare phenomenon on the solar disk. Montaigne's first three observations seemed to suggest this, but the fourth observation contradicted expectations.

During the transit, several astronomers searched for the supposed companion without discovering it. The observations of this satellite from 1764, made by six observers, amounted to seven consecutive observations and were consistent with those of Montaigne from 1761 by a periodic orbital period of 9, 10, or 12 days, provided that the inequalities in the true and apparent orbit of this celestial body were not taken into account. The orbital period deduced from Montaigne's observations by Baudouin could only have occurred in such a way that those inequalities were not taken into account.

A few years after Short's observations of the moon of Venus, Lalande searched for it several times in vain in Paris; Boscowich and Hell, despite their strenuous search, were also unable to find it. During the observations that P. Hell made in Vienna, he noticed that the bright glare from Venus sometimes produced a small secondary image on the eyeglass, which led him to conclude that earlier observers had been deceived by this and mistook it for a satellite of Venus. Lalande also agreed with this, and occasionally discussed this new discovery with Short during his stay in London, echoing the latter's opinion in the words: "M. Short, who spoke to Short in London in 1763, could not believe the existence of a satellite of Venus."

In the Ephemerides published by Hell in 1761, he asked astronomers to use their best telescopes to search for the satellite, which some believed they had observed, during the transit of Venus in front of the Sun. Hell attributed it to this announcement that people in France began searching for it even before the transit, and that Montaigne finally discovered it a month earlier. However, even these observations did not seem to convince the famous editor of the Vienna Ephemerides of the existence of such a moon; he remained doubtful, especially since no one claimed to have seen it during the transit.

On July 1, 1761, Father Hell wrote a secret letter to Abbé Lacaille, in which he expressed his doubts about this supposed discovery, but requested that the latter not publish any of the contents of the letter. Lacaille maintained the deepest silence on this point as well and died in March 1762. Two years later, Hell received a small package containing his secret letter translated into French and a reply from Montaigne.

This prompted the Viennese astronomer to publish a treatise in his Ephemerides of 1766 entitled "De Satellite Veneris," in which he comprehensively recounts all known observations, but also cites in great detail the possibility of considering the mere reflection of Venus's light on the telescope's eyepiece for a moon.

The optical, dioptric, and catoptric proofs presented there, however, are not derived with absolute precision, perhaps to make them more understandable to readers, for there were only reported for several cases. Judging from the content of the treatise, it is very clear that Hell was not so fortunate as to see the moon of Venus, but always considered it merely a reflection.

But one can also deduce from this the possibility that Fontana, Cassini, Short, Meier, Rödkier, Horrebow, and Montbarron also saw such a reflection through their telescopes. Incidentally, the author of these proofs admitted that he by no means intended to characterize the observations of this moon by others as illusory, but published only what he himself observed, in order to warn astronomers to be cautious in such observations.

After the first transit of Venus in front of the sun, there was still hope of seeing this moon in the second transit of 1769, but even then it was not seen and no further effort was made to find it.

Lambert says very aptly: "In addition to the means proposed by Mr. Hell for detecting optical deception, there is another one that directly concerns the satellite itself. For if a satellite is present around Venus, it must also appear as a satellite in its phenomena, and therefore its movement must be lawful and orderly. This is precisely what prompted Mr. Baudouin and the Academy of Sciences in Paris not to entirely reject Mr. Montaigne's observations, especially those of May 11, but to submit them for publication for the purpose of further observations. Mr. Montaigne saw the satellite four times in positions that are consistent with a satellite in the plane, and if they are examined more closely, as far as the not-so-precise data allow, the shape at that time, as well as the position of the satellite's orbit, and the orbital period of 11 days, emerge.

A significant confirmation of Hell's conjecture was provided by Wargentin in Stockholm, who in 1761, shortly after Montaigne's observations, sighted the moon of Venus and also believed he had discovered it; but when he rotated the telescope around its optical axis, he noticed that the supposed moon also moved around it like a spot in the ocular glass.

Lambert, however, sought to derive from Montaigne's observations the determination of the orbit of the moon of Venus and to read a special treatise on the subject to the Berlin Academy of Sciences. The discovery of this new celestial body caused such a stir that Frederick the Great suggested naming it after his learned friend Lambert. However, Lambert refused to accept the royal favor, replying: "Sire! I am not old enough to discover the satellite of Venus in the sky, nor young enough to see it on earth, and I find myself in a poor place, where I have no idea what this celestial body is, to aim for in the sky."

On the occasion of the inferior conjunction of Venus with the Sun on June 1, 1777, Lambert believed that the companion of this planet would be visible on the solar disk, for according to his calculations, it should have been seen below the center of the Sun at 3:00 p.m. on that day. Astronomers, however, searched in vain for it, and it was generally believed that the earlier observations had been deceptive, which significantly increased the likelihood of Hell's opinion. No further attempts to see this moon of Venus were made, for, as Lambert himself stated: "If one sees nothing of a satellite for several years, one easily tires of tracking the distant moon."

A calculation of the orbit of this alleged satellite was undertaken by the often-mentioned learned mathematician in Berlin, who also calculated tables for it which, although only accurate to individual degrees, could have yielded an approximate position of this moon in its orbit for any given time if the epoch, location in the moon's orbit, and orbital period had been correct. The elements were as follows:

Aphroditocentric equatorial co-ordinates of the potential point.

4 Z. 9°

Aphelium

11 „ 19

Ascending node

4 „ 29

inclination

64 degrees

Eccentricity

O.95

Average distance from the main planet 66 >,4 Venus radius Orbital period 11 days 5 hours.

The locations of the aphelium and the ascending node changed very quickly according to these tables.

Regarding the size of Venus's moon, the same scholar found that if the diameter of the Earth is assumed to be 1, then that of Venus would be 0.97, while that of its satellite would be 0.28. For our moon, the same calculation yields 0.27 in the same order; consequently, the moon of

Venus exceeds ours in size, although the difference between the two cannot be called very significant.

We will return to the determination of some of the quantities of these orbital elements later in this paper and try to determine them by other means.

The determination of the periodic orbital period of the moon Venus.

The imprecise location information of the satellite, which is based only on guesswork, does not give us any hope of calculating the elements of its orbit with any degree of approximation. The determinations that Lambert derived from his 1761 observations for the orbital period, inclination, and distance should therefore, as he himself points out, not make a great claim to accuracy, especially since the orbital period is presumably not very long, and an error in it cannot accurately determine the position of this minor planet for any given period.

The derivation of another orbital element also depends on the determined orbital period: the mean distance of the satellite from its main planet. This is therefore also the reason why Lambert's prediction of the inferior conjunction of Venus with the Sun on June 1, 1777, could not lead to the observation of Venus' moon on the Sun, and why astronomers searched for it in vain.

The exact orbital period cannot be determined from the approximate location measurements, but if one uses a graphical design of the observed locations of this minor planet and places them around its main planet, then the observations of 1764 show that the observation of March 15th, with respect to the same planet, occupies almost the same position as that of March 3rd, which would give an orbital period of 12 days. This determination, however, would be only very approximate, and a small error in this calculation, when calculated differently over several years, would greatly increase the period. Judging from another observation of this minor planet, to be mentioned later, it appears that in its ascending node it is often very brilliant, while in its presumably descending node it is perceived only as a dark shape.

This variation is reminiscent of W. Herschel's investigation of the satellites of Saturn, which was already reported on page 50. The current list of observations from 1764 also provides some confirmation of this, for on March 3, the satellite was near one of its nodes, presumably the ascending one, and was at that time at a great distance from Earth.

The first observer, Rödkier, remarked in reference to the description of its shape: "Weak light but distinct." Based on the approximate orbital period of 12 days just mentioned, it follows that after 6 days the satellite should be in the opposite node of its orbit. At this time, Horrebow described the shape of this moon as "doubtful," and later Lambert remarked regarding this observation: "The observation of March 10, 1764, is stated by Horrebow himself as such, from which he could not draw any definite conclusions. In fact, it, like that of March 11, fits very little with the others." Of the last observation, the same observer remarks: “The figure was small and very pale.”

Although at present nothing can be determined with certainty from a very small number of observations, I nevertheless thought, since there are no other clues, that the great variation in brightness should be taken into special consideration. Assuming that the satellite was always near the ascending node of its orbit during its first appearance in different years, one could determine an average period of these phenomena using the same method that Wurm used in earlier times to determine those of the variable stars Algol, Antinous, and others.

The period is equal to the orbital period of Venus' moon. But I cannot find agreement among all time intervals, then of course the above-mentioned assumption is not permissible and the orbital period cannot be determined in this way, but of the observations of the moon of Venus.

The average number resulting from 15,540 revolutions, which is so regularly obtained here, clearly demonstrates that the observations are not illusory or secondary images, since otherwise they would deviate significantly from the individual quantities. The period resulting from these observations appears to be correct to the second decimal place.

By adding the observation of 1759, a second determination of the same period results, which does not differ much from the one just reported. However, we shall note in the course of this essay that some other phenomena of the planet Venus can be explained by the assumption of a moon of Venus.

The observation from 1759 was taken into account; the average value found agrees with the earlier one and, as such, results from 5,563 revolutions. However, the individual numerical values of the previous calculation seem to agree better with each other, since the first two decimal places are the same everywhere. Taking the arithmetic mean of both, the orbital period of Venus' moon is 12.1 days, a value that is 2.5 times smaller than that of our moon.

The results obtained so far, which have emerged from the previous investigations, prove that here we are dealing with observations which correspond to a celestial body of our Solar system, whose existence has been widely disputed. However, by assuming this, some phenomena on Venus can be explained, which even today have been interpreted in one way or another by several scholars. Several partly self-contradictory assumptions regarding their proper interpretation are eliminated, and the phenomena themselves undeniably indicate that the planet in question is orbited by a satellite whose physical composition, with regard to its atmosphere and surface, must be significantly different from that of ours. The phenomena related to the above are the following:

1. The observed spots of Venus.
2. The striking appearances of an unusual
   shine of the same.
3. Appearances during the occultations of Venus from
   World.
4. The people who did not notice during the two passes
   to interpret phenomena.
5. The illumination of the night side of Venus.
6. The different rotation times of the same planet,
according to the observations of Bianchini, Schröter
and de Vico.

These various phenomena of Venus are so closely related to one another that they can be easily combined with the assumption of a moon of Venus and contribute even more to confirming the correctness of all observations.

1. The observed spots on the surface of Venus.

Since the invention of the telescope, the sun, planets, and minor planets have long been the subject of astronomical observations, including their physical constitution, which have provided us with the most remarkable insights into them. When Schein discovered sunspots and reported this to his prior at the monastery, the latter famously remarked: "I haven't read anything about this in my Aristotle; the cause of your discovery lies in the spotty lenses of your telescope." But several others confirmed the clergyman of Ingolstadt's observations.

THE SPOTS ON THE SURFACE OF VENUS

It once again became very difficult for scholars to explain the origin of the sun's spots. Among the many hypotheses put forward on this subject, Gascoigne's is undoubtedly the most remarkable. He believed that a large number of smaller planets orbited the sun, and that the passage of several of these at the same time created sunspots with their multiple gradations.

When a correct understanding of the sun's atmosphere was later obtained and the formation of the spots on the great sphere of light was properly interpreted, it could still be said that the passage of the lower planets, Mercury and Venus, in front of the sun's disk caused spots to form on the sun's disk, but these had a circular, jet-black shape. The discovery of the spots on Jupiter made it easier to explain them than those on the sun by drawing an analogy between the two.

However, as previously reported on page 44, dark, misshapen spots also form on this planet. However, according to page 47 of the same section, bright, misshapen spots also occur, resulting from the passage of Jupiter's moons across the planet's disk. Transits of Saturn's satellites across its disk have also been observed, as we reported in more detail on page 51; they too formed formless spots on it. If nothing had been known about the moons of Jupiter or Saturn, how could one explain the temporary appearance and disappearance of these spots, probably due to sudden changes in the atmospheres of these planets, since their origin must have been sought significantly higher than these atmospheres?

If one is truly convinced that the lowest planet nearest to us has a companion that moves around it at a definite orbital period, one must also assume that this companion occasionally passes in front of the disk of Venus, as just mentioned with regard to the moons of Jupiter.

However, the prediction of the position of these small satellites, thanks to the ingenious development of the perturbation theory of the famous Laplace and the precise tables of these celestial bodies by Delambre and Damoiseau, is currently in such good agreement with observations that the time of such transits can be calculated with great accuracy.

Most observers of this phenomenon notice that the satellite disappears in front of Jupiter. Luminous intensity that is exactly equal to the same-sized part of the planet's disk. However, some observant astronomers, whose names were mentioned in the relevant section, observed rather large spots on the surface of the satellites during such a transit. These spots alone become visible only on Jupiter, while the edges of the smaller disk disappear completely. This observation has been confirmed by recent observations, which have also shown that the rotation period of the minor planets around their axes is not always, as is the case with our moon, equal to the orbital period around their main planet.

Applying these observations to the moon of Venus, one is tempted to consider several conspicuous spots and bright points of light, occasionally seen on the surface of this lower planet, as actual transits of that companion, and to investigate whether they are connected with the orbital period determined above. The most attentive observers of Venus, besides Fontana and Cassini, were Bianchini and Schröter; the former shared his excellent observations in the work "Hesperi et Phosphori nova Phaenomena, Romae 1728," the latter in his "Aphroditographic Fragments for a Precise Knowledge of the Planet Venus, Helmstädt 1796." Furthermore, scattered reports on this subject from other astronomers can be found in the Hist de l'Academie Royale d. S. de Paris, Connaissance des tems, and Bode's Astronomical Yearbooks.

The first spots on Venus were undoubtedly seen by Fontana before Cassini, who is generally considered the first to discover them. Bianchini observed them for three consecutive years, but they were observed for the longest time, continuously for 15 years, under various alternations, by Schröter. W. Herschel also observed, with his large telescope, such phenomena as those mentioned in Transactions phil. LXXXIII of 1793. Although he had been occupied with these observations since 1777, his other scientific work did not allow him to devote much time and effort to this subject.

Each of these astronomers, however, had more or less the intention of investigating the physical structure of the surface of this planet, which often appears to us in brilliant light, while Schröter also intended to study its atmosphere, rotation, and the height of its mountains.

It would exceed the limits of this work to list here all the phenomena of this kind, especially those observed by Schröter, but it is sufficient to highlight the outstanding bright and dark spots, as well as other striking protrusions and depressions at the edge of illumination, and to record them according to the date of observation.

The assumption of a companion orbiting Venus also explains why Lamont in Bogenhausen in 1836, using the excellent refractor, did not observe any spots, although the famous astronomer carefully searched for them. In the same year, Beer and Mädler in Berlin also observed Venus very attentively several times; these observers, too, could only conclude what Lamont cited as a result of their efforts.

If we assume that conspicuous spots and bright points on the disk of Venus are created by the passage of the satellite in front of it and disappear again after a few hours, they can only be observed in those years when the nodal line of the Venusian lunar cock forms a line with the vision radius of the observer, since then the supposed moon passes over Venus in the form of a bright or dark, usually shapeless spot.

The assumption of a companion to Venus also shows that de Vico in Rome in 1841 saw the spots again in exactly the same shape as Bianchini had observed and drawn them in 1726. For while previously, and even up to now, they were considered to belong to the sphere of Venus itself, which became visible to us due to their low or high intensity of light through the atmosphere of Venus, which was very thin at that time, and represented elevations or depressions on the surface of this planet, they could not show the same shapes after such a long period of time due to such random atmospheric phenomena.

W. Herschel always argued against Schröter that the spots depended solely on the atmosphere of that planet. He never saw bright spots on the sphere of Venus and was not inclined to accept the high mountains that Schröter observed on the southern half of Venus and whose heights he measured as the result of observations.

Even the heights of the mountains of this planet, four or more miles, which the calculations made by the astronomer Lilienthal had to urge caution, as even with good telescopes they would occasionally appear as small protrusions on the edge of Venus, but this was never the case. However, if we assume these points of light or shadow to be caused by the satellite, those illusory mountain peaks disappear, and we conclude that Venus has a uniform, dense atmosphere.

According to Schröter, the most remarkable light points with the heights derived from them were the following:

1789

1790

1793

1794

The astronomer Lilienthal, highly acclaimed for his investigations into the physical structure of most of the planets and minor planets in our solar system, considered these bright spots, which he had observed on the night side of Venus, to be mountain peaks still illuminated by the sun, which had already set for this part of the planet. Based on this assumption, he was able to determine the heights of the so-called mountains. From this, this observer believed he had discovered an analogy between the physical shapes of the surfaces of the Earth, the Moon, and Venus, which consisted primarily in the fact that the largest and highest mountains are found in their southern hemispheres, and that the heights of the mountains on Venus are related to those of the Moon as the radius of Venus is to the radius of the Moon. However, for our Earth, the idea that the largest and highest mountains are located in the southern hemisphere has long since ceased to exist, since Mount Davalaghiri is higher than Mount Chimborazo.

As soon as these observations and their results were published in England through the Transactions phil. of 1792, W. Herschel was prompted to reply: “that he had not been granted the opportunity to make such rare discoveries, but this should not be attributed to the weak magnification of his instruments, nor to a lack of attention or zeal.”

More important, however, was Schröder’s excellent discovery of the twilight arc of the Venusian atmosphere, which was known as a dark stripe along the edge of the planet's illumination during its various phases, a finding that was also confirmed by Lilienthal. Regarding the discovery of the aforementioned mountain peaks, Arago makes the following remark: "Herschel's memoir of 1793 was, at first, a critique of living, and, in appearance, the smallest, somewhat passionate work of Schröter. On the fact, depending on the existence of the secondary light, which we could perceive more than 180 degrees from the contour of Venus, Herschel did complete justice to Lilienthal's astronomer.

Judging from a passage that appears on page 190 in the Aphroditic Fragments, Schröter seems to consider the spots to be water reservoirs, for he says: "The surface of Venus, according to the nature of its atmosphere, is likely to have a structure that is generally very different from that of the Moon, more similar to that of the Earth, and, like the Earth, has considerable patches of spots that contain soluble components for such an atmosphere." Regarding their rare visibility, he says elsewhere: "Even Cassini and Bianchini, who are well known as tireless observers, perceived very few spots in so many years of observation around 1666 and 1667, and the latter only in 1726, 1727, and 1728.

They, too, observed them only with great difficulty, and probably just as faintly and nebulously, indistinctly, and without boundaries, because otherwise the results would certainly have yielded more definite results and not such a striking contradiction. One can forgive Bianchini for considering the observed spots, as well as those on the moon, as landscape shadings and for giving them names in his celidography. Dr. Herschel also observed spots of the same nature, so the matter is beyond doubt. See his "Observations on the Planet Venus" in the Philos. Transactions of 1793."

After several years of careful research into the celestial bodies in our solar system, Schröter observed a phenomenon that Lahire had already noticed in 1700 and Flaugergues had also noticed towards the end of the previous century. This phenomenon consisted in the southern horn of the crescent of Venus often changing its shape or appearing blunt; this change occurred less frequently in the northern horn. The author of the Aphroditographical Fragments explains this as being caused by the shadows of high mountains that obscure those peaks.

Another observation by the same observer describes the illumination limits of the phases often with depressions, often with a significant protrusion in the night side of the planet; the same phenomenon was also seen by Fritsch in Quedlinburg. Schröter (Lilienthal) was also the first to observe that the widths of the phase parts were sometimes unequal; he even succeeded in determining the difference between them by micrometric measurements, so that there could be no deception here.

Another very important observation, which the same astronomer often repeated and confirmed, concerns the irregularity of the outer periphery of the illuminated limb of Venus. Huth in Frankfurt am Main also noted this, believing it to be a flattening of Venus. According to the discoverer, the shape of this planet should not be regularly rounded, but rather form a special spheroid. He makes the following remark: "Should this irregular spherical shape, observed so far only occasionally and only on the southeastern limb of Venus, actually be confirmed as an irregularity in the course of time, it would provide a remarkable parallel to what the Spanish mariners Don Alex. Malaspina, Jos. de Bastemente, Dion Galeano, and Cur de Canallos are said to have discovered on their voyages of discovery concerning the irregular shape of our Earth, according to the latest reports*" (p. 78 of the fragments).

We hereby report the most remarkable phenomena which we have mentioned so far, paying particular attention to their details, as they all relate to the moon of Venus.

List of striking phenomena on the sphere of Venus.

Time of observation

Shape of the corners or change of the & Remarks of observer

1645. December 25th evening

Dark spot on the southern horn

Illustration of the same in Biccioli's Almagest. Nov. p. 485.

Fontana

1666. October 14th evening

Particularly bright spots

Illustration in Schröter's Fragments

Cassini

1667. 28. April

1700. November 17th morning

At the southern horn a small protrusion

Hist, de l'Academ. of Paris p. 121 and p. 296.

Lahiire

1726. February 26th evening

Dark spot

Illustration in Schröter's Fragments

Bianchini

1788. 5. April evening

Astronomy Yearbook 1792, p. 254.

Bode

1788. 24. May evening

1789. December 28th, 5 p.m.

Southern horn rounded with a distant light point

Aphrodite. Scraps

Schröter

1790. February 22nd, 5:29 p.m.

Northern horn wider than the southern one, irregular phase boundary on the night side

1790. May 22, 9:00 a.m. - 2:00 p.m.

Southern horn much longer and narrower than the northern
1790. 9 March
6 p.m.
Isolated hook-shaped elevation on the southern horn
Astronomy Yearbook 1795, p. 212.
1793. 3. April
5:40 p.m.
Illumination boundary irregular, the horn was parabolically curved
Aphrodite. Scraps

Schröter
1795. March 19
9 a.m.
The illuminated hemisphere is significantly narrower to the south than to the north

1791. 27. December 7 a.m.
Southern horn rounded, flashing of a fine, shiny dot, I s / 4 to 2 pcs.
later the horn had its former shape
1799. 7. May
Afternoon 2 o'clock

Bright spot of light in the middle of the dark phase
Astronomy Yearbook 1812, p. 221.

Schröter
1799. 9 June
Noon
At the illumination border in the middle a larger protrusion, next to the light one were two dark
Astronomy Yearbook

Determination of an average period at the intervals of observation times.

Observations
Time intervals in days
The revolution and the period

1726 — 1645
29,283

1799 (May) — 1726
26,733
1799 (June) — 1645

56,049

1799 (May) — 1700

35,964

1726 — 1667

21,489

1788 (April) — 1726

22,840

1788 (May) — 1666

44,418

1799 (May) — 1790

33,610

1790 (February) — 1726

23,371

1791 — 1700

33,276

1799 (June) — 1790 (May)

33,050

1789 — 1645
52,601

1795 — 1666
46,780

1795 — 1726
25,223

1666 — 1645
75,980

1667 — 1645
77,940

1790 (May) — 1700
32,693

Judging by the resulting mean period, these various phenomena were caused by the position of Venus's satellite; the numerical magnitude found agrees with the previous two. Thus, the observations were perceptions of transits of the companion, which manifested themselves in several different forms.

Although it is not our intention to hypothesize about the physical organization of Venus' moon, Cassini's opinion of this companion is nevertheless very true.

Clearly, it would possess a surface that is mostly dark due to its spots, and some bright spots have the same luminosity or albedo as Venus. Furthermore, the same moon is enveloped in a fairly dense atmosphere, which, due to various changes occurring therein, can often make it almost or completely invisible.

If the bright side of the satellite is on the illuminated phase, it is either indistinguishable or appears very bright. However, if the companion is visible on the night side of Venus, it appears to us as a bright point or spot on the night side, provided its bright side is facing the observer.

In the illuminated phase, one or two dark spots are often found next to the bright spot. Cassini and Fritsch observed the bright spots on the illuminated phase in this way, while Harding and Schröter saw a bright spot on the night side of Venus. Undisputedly, the dark spots represent either the dark side of the satellite or its umbra on the disk of Venus.

The rounding of the horns, especially the change in the southern horn of the Venus phase, which Schröter often observed, is, according to the previous assumption, the passage of the planet's companion in front of that part of the surface; it is the dark part of the satellite sphere, which then appears to be facing us.

The irregular illumination boundary of the phase, which actually separates the night from the day side, can also be explained by the position of the satellite in front of the disk of its main planet. If the dark side of the companion is facing us in this case, we see a strong and dark depression in the otherwise regular boundary line. However, if its bright side is facing the observer, a noticeable protrusion appears, extending into the night side of Venus. These phenomena were observed by Schröter and Fritsch, who described them in the Aphrod. Fragment and in the Astronom. year of 1803.

Schröter noticed several times that the width of the southern crescent of Venus was narrower than that of the northern one. He had convinced himself of this through very careful observations, even with the help of two different telescopes, so that no deception could have occurred. This phenomenon also explains the existence of a moon of Venus, the dark part of which was located on the night side of the planet, covering that part of the southern phase and representing a narrower phase.

Schröter shares with us another remarkable observation, which led him to believe that the sphere of Venus could not be a regular body, since otherwise he could offer no explanation for this strange observation. These were the irregularities of the outer edge of Venus that he observed. A second attentive observer, Pluth, also saw exactly the same thing, so that here, too, there could be no deception. This latter scholar explained the irregularities he saw, as already reported, as a probable flattening of the planet. If, however, we assume that the edges of Venus and its companion, as seen from Earth, touched at the point where the irregularity at the edge was visible, and that the satellite was in front of the main planet, then a small part of its dark disk was sufficient to cause this phenomenon by entering the bright planetary disk.

This otherwise inexplicable irregularity of the circular edge of Venus can also be very well explained. The discoveries, remarkable for their time, of the dark and light spots on the surface of the lowest planet in our vicinity were probably the reason why, in addition to Fontana and Cassini, the excellent observers of their time, to whom we owe several of the most important insights into our solar system, several other astronomers made Venus the object of their observation.

When Lahire saw dark spots on the southern crescent of this celestial body, he considered them the shadows of high mountains, an opinion also adopted by the Paris Academy. This was one of the most remarkable observations made in the last year of the seventeenth century, for at that time the conclusion had already been drawn that the southern hemisphere of Venus displayed very high mountains. However, these observations made up to that point were isolated, and further confirmations had to follow, which could only be achieved through special, laborious observations over several years.

This was a difficult task, to the solution of which not every astronomer was willing to sacrifice the time occupied by other observations. For the attentive observer really only had to turn his telescopes exclusively to this planet. And they had to be among the most excellent of their time. This combination of the two main requirements is very rare at any time and in any region. The great astronomers Galileo, Cassini, Hevel, and Herschel would not have become known to us, despite their equally outstanding talents and tirelessness, if they had not possessed excellent telescopes and other astronomical instruments. But there were also observers who, despite their rare telescopes, made no special discoveries, to whom science owes no progress, precisely because they lacked the necessary effort and perhaps also the time required for observation.

One of the most learned men of his time was Francesco Bianchini in Rome. Equipped with extensive knowledge in various sciences, he also made precise astronomical observations, which earned him the favors of Popes Alexander VIII and Clement XI. In addition to this rare scholarship, he was also known as a talented draftsman, having drawn many ancient monuments as an antiquarian. In 1726, 1727, and 1728, he devoted himself to observations of Venus, which he made using very long Campanian telescopes with a focal length of 105 to 205 Roman palms. He published his findings in his work, *Hesperia et Phosphore nova Phaenomena*.

His investigations concerned the spots and rotation of this planet, which no astronomer had ever observed continuously for such a long period of time. He carefully mapped the spots of this celestial body based on his observations, and, as already reported, de Vico found the same shapes and forms in his observations in 1841. We will have further opportunity to report on the rotation period of Venus he discovered, as well as on his determination of the position of this planet's north pole above the plane of the ecliptic.

Few observers can boast of having made such precise and careful observations as Bianchini, whose precious observations will retain their value for all time. For a long time, Francesco Bianchini's scholarly work on the observation of the spots of Venus was the only one containing the best research on this planet, until 1779, when Schröter, a district magistrate of Brunswick-Lüneburg, began observations of this planet in Lilienthal, a village on the Wörpe River not far from Bremen, using a tripod-type achromatic telescope. Although the beginning was small, it promised significantly greater successes for the benefit of science, which would sooner or later lead to significant discoveries.

He continued his observations with this instrument until 1784 without publishing much of them. However, by this time he had become so familiar with the object of his observations that he recognized the necessity of more advanced optical aids as desirable. From 1784 to 1786, he observed with a four-foot reflecting telescope, but he strove to acquire even larger instruments of this type in order to make an uninterrupted series of observations on the spots, the heights of the mountains, the atmosphere, and the physical organization of Venus. This series began in 1788, when he was equipped with two Schrader's reflectors of 7 and 13 feet, to which a fine 27-foot reflecting telescope was later added.

He published his very precise observations and micrometric measurements regarding this planet in the frequently cited work "Apliroditographic Fragments," whose observations surpass those of Bianchini not only in the uninterrupted, long-term series of observations, but also in the meticulous timing of each individual one. In 81 illustrations of the phases of Venus, the various discoveries and observations are presented through meticulous drawings, leaving nothing to be desired in their execution. This valuable work is comparable in kind to the painstaking observations of Piazzi, Lalande, Bessel, and several other astronomers on the positions of the fixed stars, for it faithfully depicts the smallest details of the planet Venus at that time. The observations begin on February 28, 1788, and end on March 19, 1795; thus, the tireless astronomer was able to perceive much that no human eye had seen before. Should, in the distant future, an observer, equipped with the same or even better instruments, turn his investigations to the same planet, he will, if possible, confirm these calculations.

NOTABLY GREATEST LIGHT OF VENUS

Although the aforementioned investigations are entirely correct from a mathematical point of view, the calculated epochs of brilliance for the greatest light do not always coincide with the observed phenomena; the deviations are sometimes even significant. Some time ago, therefore, Professor Wittstein undertook a theoretical calculation that led him to assume 35 to 38 days for the epoch of brilliance before or after the inferior conjunction. An excellent treatise on the luminous intensity of Venus can be found in the Monthly Report of the Royal Academy of Sciences in Berlin (November 1860, p. 77) by Dr. Bremiker, in which he calculates it using the luminous intensity of

Vega, "Lyrae," as a unit, according to his formulas. In the course of this section, we will have an opportunity to convince ourselves whether the agreement of the observations with the aforementioned assumption reveals a slight difference or not. With this in mind, we first give, for comparison, the chronologically ordered cases in which Venus, either by day or by night, was striking not only to the astronomer, but also to the unlearned man, because of its unusual brilliance.

The first and oldest mention of such a case is found before Christ in Virgil's Aeneid, proof that this phenomenon was already known in the time of the Roman Emperor Augustus.

Since the beginning of our era, such an unusual sighting of the brightly shining Venus was first mentioned in Pliny the Elder's Historia naturalis, Libr. 2, Chapter 8. More reliably than these two, a rare, striking appearance of such a brilliance is reported by Riccioli in his Almagestum novum, p. 662, Volume I, namely, "On February 21, 1587, Venus appeared as the evening star in Italy with such an unusual brilliance that the objects her rays struck cast shadows." I have never found an earlier appearance with the exact date specified, which is why it seems to me that this is the oldest of the better known.

In February 1603, Venus appeared as an evening star with unusual brilliance, leading ignorant people to believe it was a new star. Kepler describes this phenomenon in his work "Paralipom, ad Vitell. Francof. 1604, p. 333."

In May 1609, Venus was clearly observed by day throughout France; the following year, this phenomenon was considered a harbinger of the death of Henry IV, who was assassinated on May 14.

On February 5th or January 26th in 1630, Venus was seen in Tübingen as a brilliant star during broad daylight. The then-famous Professor Wilhelm Schickard published a four-page treatise on this subject entitled: "Description of the miraculous sign seen toward the north on Monday, January 25th, 1630. Including a detailed report of the star that appeared the following Tuesday at broad noon."

It was certainly striking, however, that in that city, on one day, a beautiful aurora borealis (a miraculous sign) appeared in the northern sky, while the following day, or probably 12 to 15 hours later, Venus was seen in an unusually bright light. Therefore, the description reads:

"People chattered about a new poet's star (perhaps they wanted to say comet's star), until finally a distinguished gentleman gave them a better explanation and assured them that the bright star was often seen during the day." The author of that treatise then undertook his own calculation, which he concluded with the words: "There we have it. So boldly say that it was not a new star, but the ancient Venus."

On November 1, 1700, the day of the death of King Charles II of Spain, Venus was seen in Madrid during the daytime in an unusual splendor. The king died at 2:00 p.m. (or more accurately, at 3:00 p.m.), and thus this Venusian apparition was given a strange interpretation of the monarch's demise. In the Memoires pour servir à l'histoire d'Espagne sous le regne de Philippe V. par le Marquis de Saint Philippe, trad. de l'Espagnol. Amsterdam. 1756, T. I, pp. 54, 55, the following is reported about this event: "Then in Madrid, with attention mingled with astonishment, the star of Venus was seen shining opposite the sun. Those who were not versed in astronomy admired it as a prodigy: and flattery, still signaling itself for a corpse almost already cold, drew from it favorable omens for the eternal salvation of the late king."

The expression opposee au soleil must be subject to a different interpretation here, since Venus is never considered as a lower planet.

Sun comes into opposition. This description was given by Kästner, who very often made apt, astute remarks. Here you have the opportunity to learn about the astronomical knowledge of the Marquis de Saint-Philippe in a favorable light.The planet rose early that day as the morning star and was close to its greatest splendor. No comet was then visible in the sky and a planet other than Venus did not cause the phenomenon.

According to the Spanish original of this work, which is located at the Göttingen Library, the previous position is: „ — And on the first of November (1700) two hours after noon he expired (the King Carlos II). At that hour he saw himself, with general repair, the star of Venus shining, opposite the sun.

Those less knowledgeable in Astronomy admired it as a marvel, and even though the flattery had not yet died, the still warm corpse, faced with speculations, for the eternal happiness of the deceased King. Hailose perhaps that instant the light perished, and as far as possible from the sun, which looking at it straight on, made it shine more, that's why it seemed, and because it was declining, and with less activity the sun.

1715, June 28, 2 a.m. Venus was so bright that in France it can be easily found. This phenomenon was noticed by several renowned astronomers who were at the same time ready to observe the occultation of Venus by the Moon.

1716 on 21 July in London the daylight was shining. Venus was admired by the local people as a miracle. The phenomenon caused such a stir at that time that Halley was moved to deal with the problem of the greatest splendor.

Bruce, who was then in Abyssinia, made the report in his travelogue that the peoples there were greatly astonished by this phenomenon. In 1750, Lalande witnessed a similar event in Paris, but it is regrettable that he did not witness the same day.

In January 1776, Lambert saw Venus in Berlin as w Morning star in a very bright radiance. In the Astronomical Yearbook of 1780, p. 58, he only reports the following: "In January In 1776, I awoke one morning before daybreak and saw the shadow of the window frame on the wall of the bedroom, as if a light was shining through from afar. The severe cold at that time was terrible.

“This is the main reason why I have left it at the mere remark that Venus, as the morning star, shone in full splendor, especially since this is not unknown."

On March 16, 1777, Thierry de Menonville, while sailing on a ship in the Gulf of Mexico, saw Venus shining brightly in its full splendor, as the sun was still 5 degrees above the horizon. He describes this phenomenon in his "Traite de la Culture du Népal et de l'Education de la Cochenille précédé d'un Voyage de Guaxaca,Paris 1787, p. 47.

In the same year, in April, when Venus, as he says, was in its strongest evening radiance, Lambert again observed the appearance of a shadow cast by that planet on the window of the door between the bedroom and a second closet facing northwest. This phenomenon prompted the famous scholar to undertake an investigation into the greatest light of Venus, the results of which we have previously reported.

In 1782, at the beginning of June, Wurm saw Venus with his naked eyes at Gruibingen in Württemberg at the moment the sun rose.

In 1790 in February the same planet was observed according to Lalande's communications in Paris in broad daylight. 1793 in April .

In 1798, at the end of January and beginning of February, Venus was seen in Paris in extraordinary splendor, causing considerable excitement among the locals. Wurm remarked: "It is strange that this phenomenon could spread fear in Paris, of all places, since Lalande, in his Ephemerides for the beginning of February of that same year, had explicitly predicted the planet's greatest splendor, for he states on page 135: 'Venus will be visible in full daylight after midnight.' "

We learn the real reason for this from a letter that Lalande wrote to Bar. von Zach on the 7th of Pluviose An. VI. (January 26, 1798), in which he says: "For several days now, I have been tremendously plagued and inundated with letters and visits because of a comet which I am supposed to have announced would be dangerous to our Earth*." Something similar happened in 1872 with Prof. Plantamour in Geneva.

Some local newspaper writers may have been joking about it, but it caused an incomprehensible shock like in 1778. Dr. Burkhardt wrote even more extensively about this from Paris on February 17, 1798: "Lalande's story with the comet is as amusing as it is unpleasant for him. A joker in the Journal de l'Indicateur had the mischief of entertaining the public with the news of two comets, one of fire, the other of water, which would soon appear or had already appeared, and added at the end that the famous astronomer Lalande would probably report more details about this dubious matter to the public.

A few days later, many visits and letters came to Lalande, demanding news of this comet, partly out of curiosity and even more out of fear. He was now forced, in order to free himself from this tiresome correspondence, to have a message to the public published in the Journal de Paris; but the fear was extraordinary; a panic had seized the Parisians and showed their astronomical knowledge in no favorable light.

They ran to the Observatoire nationale to make inquiries. They heard the most absurd reports about this comet, which were proclaimed in the streets. On the very day the comet was supposed to appear and bring about the end of the world, curious people were on the Pont Neuf and the quays to marvel at Venus and Jupiter. When the supposed danger was over, the Poissardes cursed the astronomer, who, in their opinion, had caused them this unnecessary fear. Since then, this ridiculous fear has been presented at the Vaudeville Theatre and at the Citoyenne Montensier Theatre; the first received much acclaim and, considered the work of two days, is really good and funny enough; de Lalande's letter, which he had inserted into the Journal de Paris, is read verbatim at the theatre.

A treatise by Lalande G from 1773, "Réflexions sur les Comètes qui peuvent approcher de Ja Terre," caused panic throughout Paris. To dispel this fear, Busejour wrote his own work, "Essai sur les Cometes." German newspapers also tried to spread this empty rumor at the time, which also worried quite a few people. It seems, therefore, that a similar report deceives the gullible several times every century.

If one considers the striking phenomena of Venus's greatest brightness seen here over a period of 200 years, one may be somewhat surprised at their small number. The striking phenomena of Venus are explained by the fact that during the epoch of this planet's greatest brightness, a particularly purified atmosphere is capable of producing this strong light. We do not deny that in some cases this may occur with a weaker intensity of light, but very often these phenomena are indicated at a time that deviates considerably from the calculated epoch of Venus's greatest brightness.

Therefore, Wurm, who published such excellent investigations and tables for calculating the greatest brightness of Venus (von Zach's Geograph. Epliemeriden Vol. II, p. 305. Astron. Jahrbuch 1802, p. 183), says about this subject: "According to my further investigations, one may without hesitation calculate 7 to 8 weeks before the theoretically greatest brightness of the evening star, which the table indicates, and 3 weeks after it, as well as 3 weeks before the greatest brightness of the morning star and 7 to 8 weeks after it, as the entire period during which Venus, even if not at the maximum of her brightness, still has an extraordinarily advantageous position for her visibility." Now, as previously stated, 70 days must be reckoned from the greatest eastern eclipse to the inferior conjunction of Venus with the Sun, which also applies from the inferior conjunction to the western eclipse.

Therefore, according to the above, 11 weeks or 77 days before or after the inferior conjunction would have to be assumed as the period within which such a striking brightness of this planet is possible. However, this would contradict the actual theory, for in addition to the previously stated formulas of Kies and Lambert, there is another rule applicable to the illumination of this celestial body, namely: "The luminous intensity of Venus is inversely proportional to the product of the squares of the distances of Venus from the Sun and from the Earth." Lambert calculated the luminous intensity of this lower planet based on the angle it forms with the sun and Earth. He assumed this to be 1 for the upper conjunction. He then found a luminous intensity of 29,864 P for the greatest deviations, but 207,563 P for the maximum, thus almost double that for the greatest digressions. Without a telescope, Kepler's teacher Mästlin saw little.

About 300 years ago, Venus, with a deviation of 30 degrees during the day, was so well observed that he was able to measure its meridian altitude, and when it was near its greatest deviation, he often saw it for several hours during the day (Disputatio de passionibus planetarum, thes. 60, p. 35). Hagacius ah Ilayk reports in his Dialexis de nov. et prius incognitae Stellae inusit. appar. etc., Francfort 1574, p. 29, several apparitions of Venus during the day.

Although such remarkable periods of brilliance for Venus are known, which possessed a strikingly bright light, few may have noticed that the same planet sometimes loses some of its light or dims for a short time during its greatest brilliance, until it regains its previous luminosity. Kepler seems to have been the first to notice this, for he states that Venus loses its brilliance in an instant, which is probably due to the planet's large spots on its surface, and that such a dimming of light can occur due to its rotation around its axis.

The relevant passage is in his dioptric p. 81: "Accident nunc mea experimento de alterabili Veneris lumine ad noctum oculi; quae in Astronomiae parte optica recensui: Ratio nihil aliud colligere poterit, nisi hoc, Veneris stellam rapidissimo gyratione circa suum axem convolvi, differentes suae superficiei parts et luminis solaris minus magisque receptivas alias post alias explicantem."

These observations of the great man are fully confirmed by an equally attentive observer, Fritsch, in Quedlinburg in 1804. He wrote to Bode: "I have often observed Venus with the naked eye at all hours of the day. Since it sometimes appeared fainter to me in the clearest sky, and the telescope confirmed this phenomenon, I conclude that it may be subject to a random change in light. Towards the edge, a dark, curved stripe usually appears, which appears to undergo no change due to rotation." Astronomy Journal 1807, p. 205.

It is known from Kepler's reports that in earlier centuries, although very rarely, one planet occulted another, causing a significant, striking change in the light of one or both. Thus:

1563 Jupiter Saturn,
1591 on January 9, Mars on Jupiter,
1590 on October 3, Venus on Mars.
1599 on June 8, Venus on Mercury.

Wolf based his proof of the correctness of the Copernican system on this, but later, in the Latin text of Astronomiae, he only cited the second and third occultations; the others were probably only very close encounters. These phenomena must have been striking in every respect, because not only the color but also the brilliance of these planets changed.

As follows from the phenomena of the spheres on Venus we have reported, according to Cassini's hypothesis already cited, the surface of the moon of Venus is perhaps half dark, but its bright part shines with the same intensity as Venus, or its albedo is equal to that of its main planet. When this planet is at its greatest luminosity, the brightness of the moon of Venus is also at its greatest at the same time.

However, its light can also have an even greater intensity when the satellite is in one of its nodes on the horoscope of Venus, in front of its main planet. If it then assumes a position in its own orbit such that it is at its greatest distance from Venus or closest to Earth, its light will undoubtedly appear brighter to us. However, if it is at that point with its dark side facing us, we see a brief dimming of the bright light of Venus, as Kepler and Fritsch observed.

Assuming the existence of the moon of Venus, we will now compare the rare observed epochs of brilliance, whose timing is precisely specified, according to the theories of Kies, Lambert, and Wittstein, and address the discrepancy between observation and theory. According to the above, the Venusian brilliance phenomena occurred on February 5, 1630, November 1, 1700, July 21, 1716, April 10, 1777, and January 31, 1798.

Even centuries later, his observations would still be found to be consistent with Schröter's. The Aphroditographic Fragments undoubtedly constitute the main work of Schröter's excellent scholarly work. Although his observations have often not received the recognition they deserve, this was mostly based on the conclusions he occasionally drew from them. The observations themselves stand as an imperishable monument to German efforts toward scientific progress and to a rare perseverance.

Therefore, von Zach also says very aptly: "Herschel's and Schröter's names will shine in the sky like Castor and Pollux, as long as stars sparkle in the firmament, as long as posterity does not sink back to the lowest level of humanity and no longer honors that which constitutes its highest dignity."

Very often, we see the larger planets in the sky, visible to the naked eye, in a more or less bright light, which not only equals that of the first-magnitude stars, but sometimes even surpasses it. For the upper planets, Mars, Jupiter, and Saturn, the time of their most striking luminosity is when they are closest to Earth or, at night, when they appear at the meridian, that is, when they are at their opposition.

But the lower planets, Mercury and Venus, also appear in their greatest light when they are at certain points in their orbits. This is especially true of Venus, which then surpasses all the stars in the firmament and appears as a new star to those who are unaware of this change in light and have never occupied themselves much with astronomical objects.

The greatest poets of antiquity and several of modern times have praised the beauty of the star Venus in excellent verses, but less well known is that the immortal Kepler was also captivated by this sight. But the time of its greatest brilliance and at what point in its orbit Venus would then be located were not as easy to determine as with the upper planets. A special phenomenon of this kind prompted one of the most brilliant astronomers of the last century to undertake an investigation into the matter, and we owe it to him for the first time to gain more detailed insights into the matter.

When, on July 21, 1716, Venus aroused general admiration in London with her unusual brilliance, and even astonishment and fear among the common people, the famous Halley undertook to solve this problem. His treatise on this subject appeared in Transact. Philos. No. 349, 1716, p. 466, and testifies to the author's insight.

For if we assume that after the lapse of 583 days, 22 hours, and 6 minutes, Venus will again be visible to us at the same point in the sky, or that she will complete her synodic orbit during this time, then all of her phenomena must take place within this time.

The planet appears in its superior conjunction with the sun, then enters its greatest eastern digression, or eversion, then shines with the greatest light, and reaches its inferior conjunction with the sun, where it is closest to us but is not visible because it presents its dark side to us. From here, it moves further in its orbit, once again appearing in the greatest light, until it reaches its greatest western digression, or eversion, and finally returns to its superior conjunction with the sun, where it appears to us with its smallest apparent diameter, yet as a round, bright disk. The point of greatest light must therefore be in a position relative to the earth and sun that offers the greatest radiance for visibility.

After Halley, other scholars also tried to solve the problem of the greatest radiance of Venus, namely Kies, a Württemberg astronomer, and the great Euler, whose investigation can be found in the Memoires de l'Academie de Prusse pour 1750. About 30 years later, Lambert in Berlin also dealt with this subject and provided a treatise "On the Splendor of Venus" in the Astronomical Journal of 1780, p. 58*).

If one calculates the greatest deflection or digression of Venus from the Sun, as observed from Earth, one finds it to be 46° 19' 49". However, for the point on Venus where it shines most brightly, the deflection, according to Kies, is 39° 6' to 40° 21', or on average 39° 33'. Lambert, on the other hand, finds a large angle for this deflection, for he assumes it to be 44° 37' 36", if the distance of Venus from Earth is 0.53928, assuming the mean distances of Earth and Venus from the Sun to be 1 and 0.72333. According to these last calculations, the following determinations are found, expressed in days, hours, and minutes:

1) From the ongoing conjunction of Venus with the Sun to the greatest eastern eclipse 221 days, 3 hours, 5 minutes.

2) From the greatest eastern deviation to the greatest light of the evening star 19 days, 14 hours, 30 minutes.

3) From the greatest light of the evening star to the inferior conjunction 51 days, 5 hours, 29 minutes.

4) From the inferior conjunction to the greatest light of the morning star 51 days, 5 hours, 29 minutes.

5) From the greatest light of the morning star to the western greatest deviation 19 days, 14 hours, 30 minutes.

6) From the greatest western deviation to the upper conjunction of Ares with the sun 221 days, 3 hours, 5 minutes.

Lambert found Venus's greatest luminosity to be about 3,000 times fainter than the light of the full Moon. From this it seems clear that the striking epochs of Venus's brightness did not occur at one and the same node of the satellite; this does not indicate that its rotation corresponds to its periodic orbit, as is the case with most other satellites.

However, since these times do not deviate very much from the calculated maximum luminous phase, the above-expressed opinion becomes much more likely, namely that such phenomena occurred when the satellite was at one of its nodes in front of the disk of Venus around the time of the brightness epoch and increased that of its main planet through its luminous intensity.

OCCULTATIONS OF VENUS BY THE MOON

Often there was no actual occultation, but merely a very close encounter of the planet with the moon. However, more frequently it happened that ecliptic phenomena of this kind could not be observed in Europe, but only in other parts of the world where no observatories exist, making their observations impossible. Likewise, one can assume that a large proportion of these rare occultations were lost due to cloudy weather, thus resulting in the loss of even the rarest observations. Only in this way can it be explained how only a few of the occultations of Venus by the moon are recorded in the annals of astronomy as actually observed.

No such reports of the meeting of the most brilliant planet in the sky with the Moon have been found from the earliest centuries before our era, although accounts of less significant eclipses of this kind do occasionally appear in ancient writers. It is regrettable, however, that the circumstances that constitute the essence of the matter are often completely missing. It is then often difficult to separate the truth of the report from other deceptions and, at times, from other strange descriptions. Thus, Aristotle reports the following about an occultation of the moon: "The moon, when it was divided into two parts, was transformed from the obscure, and from the luminous part, it became luceret, and from the celestial part it became luminous, and from the celestial part it became lucida" (De Coelo, Book II,

Chapter 12). Aristotle cites this observation as an eyewitness, without specifying the time of its occurrence.

It is easy to understand how difficult it is to learn anything more from such a statement, since nothing is mentioned that could serve as any clue. Only a rare man like Kepler was able to determine the exact time of this phenomenon through his calculations, for he found that this occultation occurred in the third year of the 105th Olympiad, or in the year 357, on the night of April 4 BC. From these dates, it can be seen that Aristotle was then 21 years old and, according to Laertius's account, who was listening to Eudoxus.

One finds the strangest distortions in history, for how often has Venus not been described at the time of her greatest light as a comet, which until then had not been known. If the Moon had covered such a new comet, various historians would have reported it, and one should also find such events recorded in earlier chronicles. But even from these reports, the astronomer can find nothing for his science, for most of it is distorted by superstition and a penchant for the miraculous.

Two reports from earlier centuries have come to light that undoubtedly indicate planetary occultations by the

Moon. However, if the Earth's satellite had eclipsed fixed stars at that time, it could only have been Aldebaran, Regulus, Spica, or Antares. But it is even more likely that Venus was eclipsed at that time.

The well-known historian and Benedictine Aimonius reports in his History of France from the year 583 AD: "A star was seen in the middle of the Moon in full splendor: Stella in medio Lunae fulgans visa est" (Book III, 23. Chapter). The Polish cometographer of the sixteenth century, Stanislaus Lubienitz, who collected and described all possible reports and predictions about the most miraculous phenomena in the firmament in his *The Cometicum,* considers this phenomenon to be nothing other than that of a comet; another author pro coineta aereo sub Luna accenso.

A friend of Lubienitz, on the other hand, considers this story a fable, in which one would still accuse him if this subject did not merit a detailed investigation, except that instead of "the center of the moon" one would substitute the edge of the moon. For this small difference probably did not seem of great importance to Aimonius in his description, and he used a certain poetic license.

A second account of such an apparition can be found in Marcus Frytschius Laubanus's "Catalogo Prodigiorum, Miracalorum atque Ostentorum tarn in coelo, quam in terra, Norimbergae 1653." The author reports a special miracle: in the year 654 AD, a star was seen gradually approaching the moon, uniting with it, and thus forming, as it were, a single body. "Quaedam stella, contra lunam veniens conjuncta est illi, et quasi unum Corpus effectum." Here, too, it would be difficult to identify that star, for one would have to conduct an experiment with Venus and the visible planets, including Mars.

If one were to examine Jupiter and Saturn to determine whether one of these stars was meant, one would not achieve a favorable result. The larger fixed stars would still remain, which could also decide on this quaedam stella. Anyone familiar with the laborious task of a single occultation, which belongs to a distant time, will probably refrain from deciding whether that star was Venus or another planet. Therefore, no one will certainly attempt to investigate the details of this phenomenon.

For our discussion of the existence of a moon of Venus, only those occultations that occurred after the invention of the telescope are of any value. For if this particular moon were in front of its main planet at the time of the occultation, then, as already stated in the previous section, if it turned its bright side toward us, it would also lend more light to Venus, thus causing a striking phenomenon of the same kind. However, it would differ from our moon in terms of its atmosphere if we also assume, according to Cassini's opinion, that this moon has a dense atmosphere.

Under this assumption, during Venus's occultations, the light rays from Venus, which is located behind our moon, should become noticeable as they pass through that atmosphere. If the satellite turned its dark half toward us, the phenomenon would remain the same, although Venus might not shine as brightly as previously mentioned. If, during the entry or exit of such an occultation, the two edges of the satellite and the Earth's moon come into contact, the light rays emanating from Venus, which is located behind both, will refract in the atmosphere of one or the other celestial body, causing the light ray to be broken down into its seven elementary colors.

However, since the moon, as is well known, has no atmosphere, if a play of colors becomes visible during occultations of Venus at the moment the edges touch, this must only come from a celestial body in front of that planet, which has a dense atmosphere similar to ours. Let us therefore try to determine from the occultations of Venus known up to the end of the last century to highlight those in whom any striking phenomenon was noticed.

The great discoverer of the true solar system, Nicolaus Copernicus, observed the first of these and describes it in his Revolutionibus Orbiurn celestium, Norimbergae 1543, Book V, Chapter 23, p. 163. Since it was observed with the naked eye and without precise timing, it has little other value. The observation was made in Frombork on March 12, 1529, one hour after sunset.

The first astronomical observation would not be made until a century and a half later, after the invention of the telescope. This was a daytime phenomenon and was observed in Paris. Dom. Cassini and Maraldi saw the emergence of Venus from behind the bright limb of the Moon: May 19, 1692, at 3 p.m.

A subsequent or third occultation of this planet was only imperfectly observed, as the Moon was at a distance of 17 degrees from the Sun. The Moon's light was so faint that, after the entry had been observed and Venus had disappeared behind it, it could not be recognized or located again, and therefore the exit could not be observed either. Bologna Manfredi and Stancari observed the entry: June 30, 1704.

The fourth occultation of Venus was more successful, as it was observed in several locations by distinguished astronomers. Dom. Cassini and Maraldi were particularly interested in whether the moon would change its shape as it approached the planet; for if it had an atmosphere, it would be very easy to observe. The astronomers, however, noticed nothing of this. Cassini, father and son, Lahire, father and son, and Maraldi observed in Paris. At another French location, in Marseille, Chazelles, Royer, and Laval saw them. The entry occurred at the dark limb of the moon, while the exit occurred after the moon had set or below the horizon. The exit, however, would have been visible in Cadiz, Dublin, and Lisbon. The phenomenon occurred: February 23, 1708, in the evening around at 7 o'clock.

The fifth eclipse of Venus by the moon was observed even better, for several astronomers participated in the observation. What was remarkable was that Venus shone with such a strong brilliance that it was impossible to see it with the naked eye, even though this phenomenon occurred at 10:20 a.m. and 2.5 days before the new moon. Cassini, Maraldi, Malezieu, Delisle the Younger, and Chevalier Louville observed in Paris, and Plantade and Clapies in Montpellier.

This occultation was also observed in Switzerland by Professor Müller at Altdorf near Lucerne. Astronomers were particularly interested in this fifth occultation, as it was believed that it would determine whether the moon was surrounded by an atmosphere. During the appearances of two total solar eclipses, that of 1706, observed by Plantade and Clapies in Montpellier, and another on May 3, 1715, which Halley carefully observed in London, a brilliant silvery ring or halo was seen around the moon when it had completely covered the sun. When the first ray of sunlight appeared at the end of the phenomenon, however, the halo also disappeared. It was almost generally agreed that this pale glow around the moon was due to its atmosphere. Only a few disagreed, although efforts were made to prove this hypothesis on optical and physical grounds. Yet neither at the entry nor at the exit of Venus from behind the moon could Cassini, Maraldi, and Malezieu observe the slightest trace of any change in shape, motion, or color.

On the other hand, Chardalou, Chev. Louville, and Delisle made a very remarkable observation. They saw how Venus, which was white and bright while still distant from the Moon, suddenly changed color for more than a minute as it approached it closely. The edge closest to the Moon appeared reddish, and the opposite edge blue. The same colors appeared in the same order upon exit—that is, the reddish rim on the Moon's side, the blue on the opposite side.

The explanation for this play of colors was given by several different scientists, for some considered it to be the diffraction of light rays, while Dom. Cassini believed it to be the prismatic refraction of light rays in the telescopes. He believed that the observers had not seen Venus in the middle of their telescopes and thus observed this strange refraction of rays at the edges of the lens. However, to many, this phenomenon did not seem to provide any evidence for the existence of a lunar atmosphere.

Chevalier Louville artificially demonstrated this phenomenon by conducting the following experiment: He filled a round glass bottle or goblet with water and placed a light behind it. Between the light and the goblet, he held a playing card in which he had cut a small hole. The light passing through this hole represented Venus or some other celestial body, while the water-filled goblet represented the atmosphere of the moon (for Louville believed he was deciding on the existence of the latter).

He then passed in front of the goblet with the light and the card at the same time. Each time he approached it, the edge of the light hole in the card, facing the center of the goblet, appeared red, and the opposite edge blue. Yet neither the shared experiences nor the mutual arguments could decide anything in this regard. Although this experiment was considered by many to be an illusion or a product of the imagination, it seems, if one assumes that of Venus's satellite instead of the lunar atmosphere, which does not exist, to represent the aforementioned phenomenon of the play of colors very clearly. The eclipse did take place: June 28, 1715 in the morning hours.

The following or sixth occultation of Venus by the Moon was only seen in Marseille by the Minorite P. Feuillee. Before the eclipse, he observed Venus on the lunar disk, a phenomenon that had also been observed during other eclipses; the observer at the time attributed this to the effect of refraction of rays. This rare eclipse of the lower planet could not be observed in Paris due to bad weather and occurred: March 5, 1720.

It was particularly remarkable that this rare phenomenon would be visible again in the same year, namely as a central occultation, after which Venus would have moved right behind the Moon. The Moon formed an extremely narrow crescent, as it was only two days old.

This occultation caused a great stir among parts of the uneducated population of Paris, and people talked about the extraordinary miracles that would follow, for the very bright Venus appeared exceptionally clearly in the middle of the thin, dimly lit limb of the moon in broad daylight. Dom. Cassini observed in Paris, and Bianchini in Rome. This time, too, Cassini did not perceive any effect of the lunar atmosphere. The eclipse occurred on December 31, 1720.

The following occultation of Venus, or the eighth known, was observed in Bologna by Eustace Manfredi on 18 September 1827.

The ninth by Prof. Heinsius and Hofr. Kästner on February 11, 1752.

The observation of the tenth eclipse of Venus, made by Cassini de Thury, Legentil, and Maral di in Paris, was exceptionally successful. These three observers also saw the aforementioned play of colors of the planet at the moment of Venus's entry and exit on July 27, 1753.

The eleventh eclipse was observed in the County of Nice at Perinaldo by Maraldi on 21 December 1772.

The twelfth was observed only by Dr. Wolf in Danzig, although it occurred at a favorable time of year, July 1, 1777.

Likewise, the thirteenth occultation was only observed in Milan by Caesar on October 5, 1782.

The fourteenth eclipse was observed in Paris by Messier and Mechain, in Marseille by St. Jacques de Sylvabelle and Bernard on 12 April 1785.

The last occultation of Venus of the last century, which is the fifteenth, was observed in many places: in Gotha by von Ende and von Zach, in Breslau by Jungnitz, in Vienna by Triesnecker, in Göttingen by Seyffer, in Coburg by Arzberger, and in Laibstadt by Kerling. In some other places, cloudy weather prevented the observation that occurred on 24 November 1799, in the morning.

The appearance of the play of colors is undoubtedly the best evidence of the presence of a companion orbiting Venus, which must be surrounded by a dense atmosphere.

For, if one remembers during the prismatic decomposition of the light ray that the widest horizontal side of the prism produces the blue, the narrowest the red color, and that the five other known colors are contained between them, then this is the same case here. Therefore, if during these occultations the light ray from Venus passes through the atmosphere of the satellite at the last contact of the edges, the blue ray must also take its path through the densest part, closest to the celestial body, and consequently it appears on that side.

The red ray, on the other hand, passes through the upper, thinner parts of the haze, which is closest to the moon at its entrance and exit. It can therefore only always remain directed towards this last part, namely our moon, which is in very good agreement with observations. For there it is expressly noted that the blue ray of the play of colors was always on the side of Venus, while the red ray was on the side of the Moon during both contacts.

Had no satellite been in front of the disk of Venus at the same moment, these two extraordinary observations would have been in no way different from the others, since it is known and proven that our Moon is not surrounded by an atmosphere, which is also evident from these other occultations of Venus.

Only the fortuitous position of the moon of Venus at the time of those occultations, at the node of its orbit with that of the main planet, caused this remarkable play of colors.

The two remarkable eclipses of Venus from the moon, namely those of 1715 and 1753, also give cause for the following mention. During the first, which occurred on June 28, Venus had such a strikingly bright light that it was marveled at in Paris as an extraordinary miracle.

However, it was already far past its epoch of greatest brightness, which, according to Lambert's formula, occurred on March 15, 1715. According to Kies' theory, this epoch occurred on March 20, so it was seen in a very bright radiance 100 days, or almost 15 weeks, after the epoch. During this time, the moon formed a narrow crescent, for after 2 days the new moon occurred, and Venus was the morning star.

The second apparition took place on July 27, 1753, but although two days later the period of greatest light was, namely July 29, no bright radiance of this planet was noticeable, the Parisians, at the sight of this rare apparition, did not speak of imminent miraculous events, for on this day too the crescent moon was even narrower than during the earlier occultation, for after two days the new moon occurred.

The reason for Venus's dim brightness was probably that the planet's satellite, while in front of the Earth, turned its dark side toward us, and the same phenomenon occurred that Kepler first observed and which we already mentioned on page 118. Venus was dark during this eclipse and remained so until its end. It only remains to be hoped that the duration of each occultation of this planet by the moon had been accurately recorded; for the last one in the last century, according to observations in Gotha, it was 59 minutes 35 seconds.

If one seeks an intermediate period from these two phenomena, it corresponds exactly to those previously found, as follows. It will therefore not be denied that the two occultations of Venus are connected with the striking brilliance of this planet.

Geocentric longitude of Venus on June 28, 1715: 2Z. 4°

On 27 July 1753: 2 Z. 18°

Aphroditocentric longitude of the planet on June 28, 1715: 8Z. 4°

On July 27, 1753: 8Z. 18°

The difference between these two positions, if it were equal to that of the moon in its orbit, would be only 14°, which this satellite travels in approximately 13.5 hours. It therefore does not appear that it possesses the same luminous intensity at the same points; consequently, the rotation of the moon of Venus would not equal that of its orbit around the main planet. Similar observations have been made with other satellites, as already reported.

The transits of Venus across the solar disk

Careful observations of several solar eclipses have drawn our attention to a phenomenon that was not observed as frequently in past centuries as it is now, since improved telescopes allow us to notice every prominent anomaly. In particular, total and annular solar eclipses have presented observers with a question about a phenomenon that has not yet been adequately answered, and whose solution seems likely to be reserved for a later time. This strange and striking phenomenon consists in the following:

1. When the two edges of the Moon and the Sun touch, a fine thread of light first forms between them, which initially separates into several smaller sections, between which dark bands become visible.

1715, April 22 (total somnolent eclipse), observed by Halley.
1806, June 16 (total solar eclipse), observed by Ellicott and Ferne.

1843, December 21 (total solar eclipse), observed by Caldecott and Parrat.

1736, February 7 (annular solar eclipse), observed by Maclaurin.

1748, July 14 (annular solar eclipse), observed by Irwin and Short.

1791, April 3 (annular solar eclipse), observed by Webber.

1820, September 7 (annular solar eclipse), observed by Zach von S w i n d e n, and Nicolai and Schwert.

1836 May 15 (annular solar eclipse), observed by Baily and Henderson.

However, this phenomenon was not noticed:
1724, May 22 (total solar eclipse), observed by Cassini, but his co-observer Delisle saw something similar. 1842, July 8 (total solar eclipse), observed by none of the astronomers.

2. When the sun is behind the moon or very close to it, individual points appear on the moon's edge in a violet hue. These violet points of light on the moon's edge were observed:

1733 May 2nd (total solar eclipse) by Vassenius.

1737, 18 February (annular solar eclipse) by Maclaurin.

1748 July 14 (annular solar eclipse) by Short.

1820 “ 7 September (annular solar eclipse) by Van Swinds.

1836 May 15th (annular solar eclipse) by Bes. 1842 July 8th (annular solar eclipse) by Airy, Schuhmacher, Valz, 0. Struve, Schidlowski, Baily.

The director of the Munich Observatory, Lamont, offers a very apt explanation of this phenomenon in his Astronomy, for he states on page 160: "In my opinion, the bright spots in total and annular eclipses have the same origin: they are strongly illuminated spots on the lunar limb, from which a cone of light emanates.

If one asks how it can happen that individual spots on the lunar limb, while not reached by the sun's rays in a straight line, nevertheless appear illuminated, I note that one must assume a double reflection. I imagine that the sunlight enters from behind through the valleys toward the lunar limb, illuminating both the sides of the valleys and the prominent mountain peaks. From the spots thus directly illuminated (which we understandably cannot see), the light is reflected further onto others that are visible to us. This presupposes a special position of the mountains and valleys, which is why the phenomenon only appears in isolated places; it also presupposes a strong reflective power of the lunar surface." or at least individual parts of it, which is entirely consistent with other known facts." •

During the last two transits of Venus across the solar disk in 1761 and 1769, a phenomenon similar to the one just mentioned was observed, except that instead of the aforementioned points of light, a black band was visible. Furthermore, this phenomenon was observed only during either the entry or exit of Venus.

When the planet passed over the limb of the sun into the sun, the luminous part of the sun exposed behind it did not immediately connect with the limb, but after a few moments, a black band appeared, connecting the limb part with Venus. The planet thus assumed an irregular shape, especially along the side of this band; some observers compared this shape to a pear, the figurative description of which very aptly conveys an idea of this strange phenomenon.

According to the report published by Pingre on the observation of the 1761 exit, he says: "At the exit of Venus in 1761, before the edges touched and where a perceptible gap was still visible, I saw a black streak from Venus extend and reach the limb of the sun; I took this moment for the inner touch. Many have seen something similar during the entry this year (1769); we too were expecting the phenomenon (namely on St. Domingo, where Pingre, with Fleurieu, Filiere, and de Touris, observed the second transit of the past century), but neither I nor my companions were able to perceive the slightest of it."

During Venus's entry into the Sun in 1769, many observers in Europe saw this phenomenon, but some did not. For example, at the Greenwich Observatory, Maskelyne saw the same phenomenon, while Horseley, who was observing there, did not.

If this phenomenon were visible at all during the transits of the lower planets, it would also have to occur during those of Mercury, but none of the observers noticed it, and it is never mentioned in any of the reports of this observation. To better confirm what has just been reported, let me quote here some of these astronomers' own words about the transits of Mercury: Paris, passage of 1799, May 7th.

Here, too, Mechain saw Mercury surrounded by a haze as it passed by; but I believe this is the fault of the telescopes, the clouds, and the vapors. Mechain saw Mercury completely clear and isolated with the telescope at

Platina, which has sharp edges; but with his achromatic telescope he observed it as if surrounded by a small mist and a round glow.

Lalande, the same passage. “The haze mentioned here may originate from the vapor glasses. Schröter in Lilienthal, Seyffer in Göttingen, and Schmidt in Darmstadt also observed the same v. Zach.”

In Krakow, when recording the moments of this passage, the following remark was made: “The thread of light between Mercury and the edge of the sun was taken for the inner contact.” Excerpt from a letter by Ed. Troughton, London, 27 May 1800."..One of my good friends, named Gregory, who may be known to you as the discoverer of a comet.”

Hardly had the clouds dispersed and the planet become visible on the solar disk when he called me and asked me to see if I also saw something similar to a penumbra around Mercury; but I saw nothing of the sort. To my eyes, the planet appeared completely dark and well-defined, without the slightest rim of light or shadow. When this friend noticed that I was receiving very strong encouragement from a curious group, he left and went to another friend who was observing this phenomenon in the Temple.

There he saw the same phenomenon around Mercury as I had. A certain Butt, who was observing two miles from the city with excellent telescopes, both achromatic and reflecting, also saw Mercury surrounded by a bright rim; his wife and daughter saw the same thing. According to Flaugergue's treatise on the passage of November 5, 1789, nothing remarkable was observed.

Nothing unusual was reported about the passage of November 8, 1802. It simply states: A. Schröter measured Mercury on the Sun and found its diameter to be 8.9 inches. The nebula ring around the planet was once again clearly observed in Lilienthal with various telescopes, including by the official auditor, Lunder, who was present at these observations. Pre-passage of 1832.

"When I observed the first transit in Danzig on May 5, 1832—none had been observed since 1802, and of that only the exit at the Paris Observatory—a local zealous friend of practical astronomy sent me the message that in his telescope he saw a small black circle on the sun's disk, which could not be the planet, since according to predictions at least half an hour would have to pass before the entry. The observation was correct; the black circle was a sunspot, which, strangely enough, appeared perfectly round on the day in question.

When Mercury later entered the orbit at the predetermined hour and minute, it differed from the only slightly smaller spot by being much darker, and I had the opportunity to convince myself for the first time that, as previous observers have claimed, sunspots are by no means black, but still reflect light, and only appear much darker than they really are due to the contrast against the bright disk of the sun. It is in the nature of things that the entry must always be seen somewhat late, since the dark planet can only show itself when a part of it is already on the disk of the sun. Nevertheless, the inner contact of its edge with the edge of the sun can be observed very clearly, as can the exit, and here also the complete disappearance.

The reports of Bessel, Harding, Gauss, Mädler, and other observers of this transit also mention no striking phenomenon, either at the entrance or exit, and it is thus established that the transits of Mercury differ in this respect from the last two transits of Venus. Let us, however, attempt to gain some insight into this strange phenomenon, for we cannot attribute it to irradiation, since it was observed only at one marginal contact and not at the other. If it were an effect of irradiation, it would have to be visible at both entrance and exit.

If, according to Lamont's very probable explanation of the origin of the points of light and the small black streaks between them during solar eclipses, we must attribute them to the uneven edge of the moon, the assumption of an irregular edge is invalidated in the case of Venus. The same planet always appears round and surrounded by a dense atmosphere, which, as W. Herschel very aptly remarks, does not allow the inequalities on its surface to be perceived. We have previously attempted to designate several significant spots on Venus, observed by Schröter, as those on the satellite disk during its passage in front of its main planet; these spots were seen in their various appearances in such a sequence of time that one could also deduce an average period from them.

Therefore, if nothing exists on the disk of Venus that could produce this rare, extraordinary formation of the planet, then there must be another cause for its occurrence. It is particularly noteworthy that during the transit of 1639, of which only the exit was observed, neither Ilorrox nor Crabtree observed this rare phenomenon, so it is not always visible. However, it does not occur during either transit, but only either during entry, as in 1769, or during exit, as in 1761. Likewise, not all observers saw it; even some of them at the same observatory noticed nothing. There was no thread of light extending across the two edges connecting them, but according to the description of some attentive astronomers, there was a dark stripe, according to others, a round shape that merged with Venus.

Afterwards, they formed a body that took on a cerebral contour, which soon disappeared, and Venus remained only as a round disk on the sun during entry, or disappeared during exit with the shape described above. If there is no adequate explanation for this phenomenon, the only obvious possibility is that it was caused by the moon of Venus.

Don't the descriptions of this phenomenon agree with the assumption of a companion, since even the dark color was in no way different from that of the planet? The only thing that seems most remarkable here is the location of this satellite, which, at the moment of contact, was next to its main planet and, by merging its small black circular area, only a quarter of the diameter of Venus, with the large one of this planet, could form a brain-shaped figure.

If we further consider that this phenomenon was not observed by astronomers from all over the world, who simultaneously observed the transit at different points on Earth, the reason for this can be attributed to the effect of the parallax of the moon of Venus, which prevented it from being observed everywhere. That the position of that small moon was next to that of the main planet during both transits was a rare coincidence, which will probably not occur during the next transit in 1874 or in 1882. We shall be even more inclined to assume this phenomenon for that of the moon of Venus if we consider more closely the observations of it by Montaigne in 1761, since they were made only four weeks before the transit of Venus.

The observations lie in such a narrow ellipse that its minor axis is almost equal to the diameter of Venus. Furthermore, according to a further calculation, Venus's companion appeared in its ascending node on May 6th; it must have been in its descending node on June 5th of the same year. If one calculates its mean position using the previously determined orbital period for June 3rd, 1769, namely the second transit, one finds that it was very close to the aforementioned node on that day as well; however, in both cases, it was further from the Earth than Venus, or it was beyond it.

There can therefore be no doubt about the identity of the mentioned phenomena with the moons of Venus. At the first transitions of the last century, we resigned the companions of the small Venus coming from its descending node above the edge of the main planet and stepped in front of it.

The reports of the second passage of 3 June 1769 agree that during the entry, Venus merged with the edge of the Sun for a few moments through a black band. Venus's companion passed behind the disk of its main planet and remained there for about 2 hours, or 56 hours according to the calculation of the mean motion, and since after the observation of the first contact for Europe the Sun soon set.

It was not noticed by the observers during its second exit over the limb of Venus on the solar disk. It could have been seen by those astronomers who were able to observe the entire transit in the far north. However, they probably did not look further two hours after observing the entry to see if anything could still be seen beneath Venus.

It is also noteworthy that they observed the first inner contact too early, namely Hell in Wardshoe by 14 seconds and P.I. Banman in Cajanneburg by 20 seconds. For if we imagine that due to the superposition of Venus and its moon on the side of the limb of the sun, the curvature of Venus acquired a different curvature, the first inner contact must naturally have been observed earlier, and P. Hell and P.I. Hanman could be justified against the accusation of inaccurate observation of this first inner contact.

The satellite then traveled the small distance from the southern limb of Venus to the nearest limb of the Sun unnoticed in its orbit and passed beyond this second limb much earlier than the second inner contact occurred, thus becoming invisible. Only in this way can the phenomenon that became visible during the transits be explained, and the calculation based on the mean motion of the moon of Venus is no different.

Soon after the observation of the last transit, a report appeared from a certain Abraham Scheuten in Crefeld. He claimed to have seen this supposed satellite in the sun for three hours on June 6, 1761, and that it was as distinct and round as Venus. However, if it was really an astronomical observation, he mistook a small, round sunspot, such as frequently appear on the sun, for the satellite, because his alleged observations do not agree with Montaigne's observations from May 1761. For a better overview, we are sharing the passage from his letter, which he addressed to Lambert in Berlin.

From the Astronomical Yearbook of 1777, I see that astronomers are still uncertain whether Venus has a satellite or not. Since I am now completely convinced of this, I share here the evidence of my conviction, because it seems important enough to me. I and my friends have often seen the satellite pass very clearly in front of the sun through a common telescope within three hours.

In my Memorial, I find the following: "In the year 1761, on the morning of June 6, at 5:15 a.m., I saw Venus in the sun. From 8 a.m. to 12 p.m., no observations could be made here because of the clouds. At 12 p.m., I saw Venus's moon in the center of the sun's disk. At 3 a.m., it was almost at its limb. "What we saw in the sun during these three hours could be nothing other than the satellite. It seemed to me as black, round, and distinct as Venus, but much smaller, about 1/4 as large. It also didn't look at all like the sunspots I've seen many times. Its orbit coincided with that of Venus; however, it was somewhat faster.

From this, I surmise that it was on the side of its orbit closest to our Earth, and thus not far from its inferior conjunction with Venus, which also caused the satellite to appear all the blacker. Due to a lack of instruments, greater precision was not possible, but it was enough to convince me of the satellite's existence. I would have made it known sooner, but I suspected that many would have seen it."

Notes by Lambert. I asked the unnamed author of this letter, on the very occasion through which his letter reached me, to provide some details of his observations, if he had them available. I then received the following

Second letter from Crefeld, December 28, 1775. "I am sorry that I cannot give a completely satisfactory answer to the questions addressed to me. If this result were only conjectured from my observation, it would have been recorded more precisely and comprehensively, as far as possible given the lack of instruments. Witnesses to the phenomenon are still alive. Without knowing anything about astronomy, they saw the moon of Venus, in its expression, like a black pea in the sun. This fact is in itself irrelevant. The following is all that I can still report about my observation.

"The first observation was at 12 o'clock or a few minutes later, and the moon, by eye, was directly in front of the center of the sun. I do not know exactly how far it was from the edge at 3 o'clock, but it was still just visible; I am therefore no longer able to give a diagram of it. I deduced the speed as follows: I divided the diameter of the sun into 100 parts. Of this, Venus moved about 80 parts in 1 hour 20 minutes, which is 12 1/2 parts in an hour. The moon moved 50 parts in 3 hours, which is 16 2/3 parts in 1 hour, and therefore faster than Venus. Abraham Scheuten, Adam's son."

Postscript by Lambert. "This observation by Mr. Scheuten now clarifies to me that I could have seized an opportunity, which actually existed at that time, to see and observe the moon in front of the sun's disk in Augsburg. On June 5th, the weather was doubtful, and the sun could not be seen before its setting. Therefore, for the following morning, there remained only an uncertain hope that the weather would clear up during the night.

I got up at 4:00 a.m., the sky was bright, and a balcony above my living room was a convenient place to see the sun immediately after its rise. At 5:00 a.m., the sun rose from behind the surrounding roofs, and Venus was exactly in the vertical line of its center, 1/2 of the diameter of the sun from the rim. Through this single observation, I was able to draw the true apparent orbit of Venus in the figure I had already drawn the day before and to plot the hour. Since it turned out that Venus would not emerge from the sun before 9:15 a.m., I had time to contact some friends, who wanted to join in, to share the news.

A little before 9 o'clock, thin clouds appeared in front of the sun, so that one could see the sun with the naked eye. Some of the spectators saw not only Venus, but also said that they saw a smaller one. At that time, I knew nothing, and replied that the smaller Venus would be a sunspot. I didn't look any further because I had already seen enough sunspots and did not know that there was still something else to look for and observe. Now I wished I certainly paid more attention to it, although, since the clouds always increased, the observations were not many.

Although it is striking that apart from Scheuten no other observers saw the moon of Venus on the disk of the sun. One cannot find any other explanation for this sighting. In the transition before 1832 there was a similar phenomenon which we have shared on page 135. Even then, a second planet or a second Mercury, which is a little smaller than what appeared and which was also a sunspot.

According to Enger's remark, it was not as dark as the one planet, which is in the deepest black in front of the bright disk of the sun. Scheuten also saw through a common telescope: the same in a significant diameter, which was not predicted,

The situation was different with Pingre's departure. The same phenomenon seen before, because, as already mentioned, it is in perfect agreement with Montaigne's observations.

According to these observations, the ascending and descending nodes of the satellite's orbit fell almost within the observer's extended field of vision, or the orbit formed such a narrow ellipse that it was almost a straight line. The position of the satellite's nodal line changed only slightly from the days of Montaigne's observations to the days of the transit; therefore, this small celestial body, being behind the disk of its main planet during the aforementioned phenomenon, could have been invisible to us, and could only emerge above the disk when the exit was near.

The remark of Lambert's friends that they had seen a smaller Venus at that time is much more important than Scheuten's perception, which here certainly does not correspond to any of the previous observations of the moon of Venus. However, If Scheuten has mistaken a sunspot for the satellite of Venus, it has probably often happened the other way around, that the satellite was mistaken for a sunspot and no further attention was paid to it. Therefore, special care must be taken with the above transits to avoid confusion.

Apart from its circular shape, a moon of Venus must make itself noticeable by its rapid movement across the sun's disk, which surpasses that of Venus, which in turn surpasses the sunspots. The satellite will probably appear in pitch black, surrounded by a ring of fog, as it has a significant atmosphere. Sunspots, however, also appear with a somewhat brighter surrounding, so confusion is very easy, and they differ only in their slower or faster movement.

During the frequently occurring inferior conjunctions of Venus with the Sun, it has probably already happened that, if their distance from each other was not very significant, the moon of Venus appeared on the Sun. Many attentive observers may have seen it without considering it anything other than a sunspot. Often, small, round spots, similar to planets, were reported, appearing on the sun's disk for a while and then disappearing. I have examined the reports of such phenomena, with approximate dates, found in the most well-known works on the subject, but I have not found a single one that could be considered with any probability to be a moon of Venus.

The illumination of the night side of Venus

Already in the Middle Ages, various astronomers observed the well-known phenomenon that our moon, a few days before or after the new moon, when it appears to us in the sky in the shape of a narrow crescent, shows us its sometimes faintly illuminated night side. The French have their own, very descriptive term for this illumination of the night side of the moon, for they call this light lumiere cendee. According to some, Michael Möstlin, who died in Tübingen in 1590 and was the teacher of the great Kepler, was the first to explain the cause of this ash-grey light, while the Italians attribute it to the famous painter Leonardo de Vinci.

The latter seems more correct, for an excellent description of this phenomenon, along with its explanation, has been found in his manuscripts, and he also lived before Möstlin, for he died in 1520. A treatise on this subject entitled: Essai sur les ouvrages physico-mathem. de

Leonardo de Vinci by Venturini, Professor of Physics at Modena, Paris in 1799, proves very thoroughly that the latter opinion is the correct one.

However, many also found the explanation given by Leonardo de Vinci or Möstlin regarding the origin of the ash-colored light on the moon's night side implausible. It is well known that at this time of year, when viewed from the moon, the Earth appears as a disk four times the diameter of the moon. It is almost fully illuminated before or after the new moon, and its intense light shines significantly brighter on the moon's nights than we perceive it on our planet, even when the moon shines as a full moon in the sky at night.

Among the opponents of this natural explanation, the most prominent was the then famous Fortunatus Lincetus, who invariably repeated and maintained that the light on the dark lunar disk was due to the moon being a large, luminous Bolognese stone, which he defended in his work "De lunae subobscura luce prope conjunctione etc. Utinae 1642 in 4to. Gassendi tried his best to dissuade him from this absurd opinion, but in vain, for Licentus went so far in his errors that he even denied the refraction of our atmosphere. However, since the invention of the telescope, it was generally believed that this reflected light came from the

Earth illuminating the dark moon. It is known that sometimes, although very rarely, soon after the inferior conjunction, when the crescent of Venus is very narrowly illuminated, the night side of this planet becomes visible to us in a special color. It is precisely this color that is probably impossible to define, but it is almost similar to the night side of our moon when it becomes visible to us.

Given the current state of science, it is still impossible to determine in advance when this phenomenon might be visible, especially since we have so far had few observations of this phenomenon. All we can say is that it is possible to observe this phenomenon several days before or after the inferior conjunction of Venus and the Sun, if the planet is very close to Earth and other causes still unknown make it possible for such illumination of the night side of Venus to become visible.

Some time ago I found in Riccioli's Almagestum novum that he was the first to discover this phenomenon, for his observation dates from January 9, 1643, and was therefore made 33 years after the invention of Galileo's telescope; it is described in the same terms as those of later observers.

The famous Jesuit and opponent of the Copernican solar system, Joseph Baptist Riccioli, was the first to observe the secondary light of Venus. He writes about it: on January 9, 1643, the tube of the Fountain was arranged for me, in that position and shape, which can be seen in the present diagram. . . It was indeed red toward the Sun, fading in the middle, and blue-green on the side facing away from the Sun; but that variety was probably from a glass tube. The bright half-ring, with which she was crowned behind, was perhaps from Jove and Saturn illuminating her, as can be the case with the eastern ones. . . Riccioli Almagestum novum Tom. I, p. 483.

As is clear from this communication, Riccioli considered the illuminated night side of Venus to be an optical reflection from his telescope and attributed its peculiar color hue to the objective glass of his instrument. Otherwise, this observation is consistent with the inferior conjunction of Venus and the Sun, which differed by only as much as in other observations of this kind.

It was generally believed that William Derham, an English clergyman, in his "Astro-Theology or a demonstration of the Being and Attributes of God from a Survey of the Heavens" first drew attention to the phenomenon of the occasionally visible night side of Venus in the first 1930s. However, the same observation seems to have been observed in Germany even before this time, although the observation was not published at the time.

It was not until 1809 that an older observation than Derliam's was discovered, and Bode published it in his Astronomical Yearbook of 1812, p. 221, where he states: "In the old Kirch diaries at our observatory, I found the following passage from Kirch, the son: Today (June 7, 1720) I thought I saw the dark side of Venus through a 16-foot (ordinary) telescope, when the planet was only very narrowly illuminated."

At the same time, Dr. Olbers also managed to find the following remarkable observation in Andr. Mayer's Observationibus Veneris, Gryphiswaldense 1762, p. 19, which he communicated to Bode: spring October 20, 0 h 44 '49' 9 " (Passage-Instrument enters Horn) decl. aust. 21° 31' exacta sereniss. It is the bright part of Venus and was very thin, nevertheless an entire disc appeared, like the waxing moon, which reflected the light received from the earth. — Das zur Beobachtung angewandte Fernrohr war ein Sechsfüssiges Passage-Instrument von Bird und die seltene Erscheinung wurde bei hellem Tage gesehen.

It is curious that the dates given in the report of these observations are incorrect, for in Kirch's observation it must be 1721, and in Andreas Mayer's 1759. A very good chronological compilation of all observations belonging to this category can be found in "H. Klein's Handbuch der Himmelskunde, Braunschweig 1869," which provides an overview of such phenomena that one will search in vain for in all works on astronomy.

The perception of the dark phase of Venus can be classified according to observations into those in which the entire night side was visible, and others in which only a part of it was visible in the light peculiar to this phenomenon. We will therefore present the observations belonging to this category here, arranged according to these subdivisions.

Derham was born in 1657 and died in 1735. The above-mentioned work, as well as his "Physico-Theology, or a demonstration of the Being and Attributes of God from his Works of Creation," were highly esteemed in his time. A translation of the former work by Fabricius appeared in 1739.

Entire night side visible

On January 9, 1643, Riccioli observed Venus in Rome as the evening star, which had an extremely narrow crescent. The part of the planet that was facing the sun or illuminated by it appeared reddish, but its center was yellow. In contrast, the part facing away from the sun was perceived as bluish-green. A narrow ring of light was also observed at this same part. The observation was made with a nine-foot Fontana telescope.

On Saturday, June 7, 1721, Christian Kirch in Berlin wrote in his diary: "I found Venus in a region where the sky was not very clear. The planet appeared narrow, and it seemed to me as if I could see its dark part, although this is almost unbelievable. The diameter of Venus was 65 inches, and its crescent seemed to tremble in the vapors."

March 8, 1726, Friday. The same observer makes the remark: "We observed Venus through the 26-foot tube. I recognized the dark side, and indeed, its edge seemed to me to belong to a smaller circle than the bright edge, similar to what is usually the case with the moon. Messrs. Harper and Möller confirmed the observation. I had the usual cover of the tube, but not a narrow opening as is usually used when observing Venus. At the boundary of light and shadow, Venus appeared darker than the outer bright edge, and as if spotty."

1759, October 20, 0:44' 47.9" true time in Greifswalde. The southern declination of Venus, which I observed, was 21° 31', and although the luminous phase of the planet was very narrow, one could nevertheless perceive its entire disk, like that of the moon when it is a narrow crescent, which reflects the light received from the earth.

On January 24, 1806, at 7:10 p.m., Harding observed Venus in Göttingen with the ten-foot Herschel reflector and 84x magnification. The observer said: "With the full aperture of this beautiful instrument, I saw the entire sphere of the planet not illuminated by the sun, which by a dull ash-green light against the dark sky - clearly distinguished itself. However, the illuminated crescent of the planet was much larger against the dark disk, as was the case with the moon, so a phenomenon which, according to optical reasons, is very easy to explain."

1806, February 28, 6:12 p.m., the same observer says: "With twilight not yet ended and the air clear, I saw with the aforementioned telescope the dark globe of Venus again—once again without a doubt. This time, however, the brighter part not so much larger than the dark one, undoubtedly because the observation on January 24th was made in complete darkness, while today's was made in twilight that had not yet ended, and moreover the bright crescent of Venus was now already much narrower than it was then.

"This time it appeared in a dull, reddish-grey light, like the moon during total eclipses. Moreover, the phenomenon was exceptionally clear, and the edge of the planet's disk was particularly sharply defined."

On March 1, 1806, when the previous observer was still observing Venus during twilight, using only the five-foot reflector, he immediately saw the night side of the planet in full clarity, and the ten-foot reflector completely confirmed this observation. Never before had he seen it so clearly as now; the outlines were extremely sharply defined and stood out so strongly against the darker blue of the sky that even the optician Gotthard, who had arrived, recognized them at first glance.

1806, February 14th, at 7 o'clock in the evening, Schroter found Venus at an altitude of 10°, the air was yawning. Nevertheless, at 150x magnification and with the excellent spoon-shaped reflector fully open, it appeared exceptionally clear in its sickle-shaped form, sharp and beautifully defined, with both horns tapering off finely and regularly.

"But without thinking about anything further," says Schröter, "it occurred to me, after so many observations of the crescent-shaped figures and atmospheric twilight of Venus, which were cited in my aphroditographic fragments and partly made with the 27 cracked reflector under a 20-inch aperture, that for the first time the entire area not illuminated by the sun was remaining a sphere, in its night form, in extremely dim dark light in my face. It was not, however, an illusion, because against the strong light of the illuminated day side, I saw it just as noticeably smaller than its circumference, as is the case with the unaided eye with the crescent-shaped moon, and because I could not remove this easily explained optical illusion, no matter how I imagined it.

On June 8, 1825, at 4 a.m., the same phenomenon was seen by Dr. Gruithuisen in Munich. It occurred 21 days after the inferior conjunction of Venus with the Sun. The planet had risen before 3 a.m., and its height above the horizon could not have been significant.

On April 20, 1865, Dr. Engelmann observed this phenomenon in Leipzig; the color tone was grayish-green compared to the background of the sky; the inferior conjunction of the planet with the sun occurred 19 days later.

The color tone of the phenomenon is described so differently by observers that it is in the interest of science to share the different names for it here:

Biccioli . . . bluish green,
Hardwig . . . ashen gray, roth gray,
Schröter . . . matt dark,
Engelmann . . grey-greenish,
Hahn . . . grey to brownish.

Judging from these descriptions, one and the same observer saw the night side of Venus in a different light on different days.

Entire night side visible during a solar eclipse

We owe a valuable information on this to Professor A. Winnecke in the Astronomische Nachrichten No. 920, p.

119, in which the relevant passage is the following: according to a kind communication from Dr. Klein, there are already 23 observations of the visible night side of Venus in the current issue of “Gaea”, No. 3, pp. 154 to 171.

"In the local (Göttingen) library, one can find some writings by Derham, in which Humboldt's note in Kosmos, Vol. III, p. 525, that he had seen small stars between the disk and ring of Saturn, is not mentioned. On the other hand, one finds in his Astrotheology that he had already known the ash-colored light of Venus. "He says there:
"... This round spherical shape is clearly visible in our moon, as well as in Venus, where, when its shape is pointed, one can even recognize the dark part of its sphere, as it is visible under a dark and inconspicuous color." (Translation by Fabricius 1739, p. 122.)

"In a note, he speaks further about this, saying that he persuaded a friend to look for the ash-grey light of Venus during a solar eclipse, and that the friend actually observed it. — In Chr. In Meier's more recent works, no observation of the night side of the sphere of Venus is mentioned, which is why I consider the observation of Derham and his friends worth sharing."

Which solar eclipse was during which the phenomenon was observed is not reported here, but this is not important, since it at least offers a refutation by proving that the visibility of the night side of Venus is due to reflected Earthlight. It would also be possible, however, to determine the time or nature of that eclipse if certain conditions related to the matter itself were introduced into the calculation.

Partial visibility of the night side

When one considers the few observations mentioned above, which have been made over the space of two centuries concerning the secondary light of Venus, one will certainly admit that this must be a very rare phenomenon. However, the observation of the moon of Venus also occurred quite frequently, although it was even more striking than the former.

However one may derive the cause, several circumstances must still cooperate simultaneously to produce its visibility. One can also add that all observers of this phenomenon discovered it only by chance and were surprised by this observation, initially believing it to be an illusion. However, if they had known in advance that the night side of Venus would be visible at a particular time, they would have observed it more often, and we would find a greater number of recorded observations. The current state of science does not yet allow us to develop sufficiently and convincingly mathematical formulas and to make the necessary calculations accordingly. In this paper, we only intend to present the causes that produce the visibility of the secondary light, and we therefore believe we can present at least some explanation analogous to that of the secondary light on our moon, which is consistent with our previous descriptions of the moon of Venus.

If an attentive and learned observer, such as F. Bianchini in Rome, at the beginning of the last century, had ceaselessly observed Venus with good telescopes, observing every rare phenomenon on its surface, and continued these observations for three years, might he not have observed a partial visibility of the planet's night side?

However, he saw nothing of this, as he meticulously recorded his observations.

Schröter observed the same planet for 15 years, and although he observed a partial illumination of its night side several times, it was only several years after the end of his observations that he managed to see the entire night side of Venus illuminated for the first time, thus considering himself the discoverer of this phenomenon, which had already been observed before him. Thus, one can judge how rare phenomena of this kind are, and that even the most tireless observer fails to perceive them. This remark refers to both the total and the partial visibility of the night side of Venus, of which we will here report what is known so far.

The first observations belonging to this area were made by the hereditary marshal von Hahn in Remplin and Schröter both published almost simultaneously, because the first wrote to Bode on May 13, 1793: "For some time now, I have often observed Venus, especially during the day. On this occasion, I made a remark about which I would like to request your opinion. Very often, I saw with great clarity the dark part of the planet's disk, which was noticeable by a grayish-brownish color. I would like to know if any false light or similar cause could cause this phenomenon, which one observes through several telescopes when the air is clear.

"This gives Venus an oval shape. At times, I even thought I saw the completely rounded disk, surrounded by a fine circle of light at its edge. Now, with the planet's more crescent-shaped shape, this is no longer so striking; instead, it is bordered at the bottom by a black stripe, as the figure shows."

This illustration shows an inverted crescent of Venus. The invisible night side of the planet is turned downward, and the concave boundary between the light and night sides contains a narrow line or stripe from one horn to the other. This is the twilight arc discovered by Schröter, from which this astronomer calculated the height of Venus's atmosphere. However, the Lilienthal observer had already seen this arc on August 12, 1790, and recognized the true cause of this phenomenon. The same observations were recently confirmed by Secchi in Rome, but the latter astronomer found the twilight arc somewhat larger than Schröter.

Von Hahn's above statement suggests that he also saw a partial illumination of the night side, since he later saw the entire disk. Incidentally, this shows that Riccioli made a very precise observation in 1643, for he already mentions the fine circle of light at the edge of Venus, which he calls the semiannulus lucidus.

On May 16, 1793, at 7:40 p.m., Schröter also made a similar observation to von Hahn, as he reported: "Venus appeared extremely narrow; I discovered an even fainter light than that of the twilight arc, this time glowing primarily and more noticeably at the southern tip of the horn, which mingled with the color of the sky itself at both tips; the light glowed primarily at the southern tip of the horn.

On May 21, 1793, at 8:30 p.m., he also observed such a light, but it glowed at the northern tip of the horn, and the bright night side of Venus appeared in a dim light. On December 23, 1794, before the setting of the sun, Venus was illuminated very narrowly, but it had a clear image and Schröter saw on both horns a spherical light of approximately the same length, reaching around and over the semicircle into the night side, a noticeably fainter, twilight light, which also fell away in a fainter way and was lost in the color of our bright twilight air.

On January 18, 1806, the night side of Venus appeared quite sharp and in some places as if finely jagged;

but from the luminous edge to two-thirds of the phase, Huth in Frankfurt am Main saw a dull, weaker grayish light, which increased within an hour and at 5:30 p.m. extended to three-quarters of the phase at the night edge, so that only a bright ring y-4 the width of the phase remained at the edge. The southern lobe, which had been pointed at 4 o'clock, was rounded and blunt, while the northern lobe remained pointed, as before.

Unequal brightness of the two horns of Venus

On May 16, 1806, Venus appeared illuminated above y3, the night boundary quite sharp, but slightly shadowed; the northern horn was slightly more curved than the southern one. It was striking that the southern half was considerably brighter than the northern one, so that while the former appeared to have reddish sunlight, the latter, according to Huth's observations, had a dull whitish moonlight.

On May 26, 1806, Venus was brightly illuminated, its nighttime boundary blurred, its edge very sharp, and its southern half brighter and fiercer than the northern half. The observer Huth further stated: "What I had never noticed before, I now noticed: the northern half was more flattened than the southern half, right in the middle. Could this have been a true flattening of Venus, and could this be its north pole? Both horns did protrude slightly, but the southern one was rounded, the northern one pointed."

All the observations just mentioned prove sufficiently that the perception of secondary light at no time of day or evening is tied to it; solar eclipses also have no influence on them, but they depend only on the time of the inferior conjunction of Venus with the Sun. Let us try to determine the difference between this and the phenomenon expressed in days, to at least convince ourselves whether they occurred more frequently before or after this conjunction:

Time of observation

Time of the lower

Conjunction

Difference in days

17 June

4- 10

1726

April

+ 27

1759

March 12

+ 47

February 14

+ 26

1825

6. May

+ 16

The difference derived from the phenomena above proves that, with the exception of one, they all occurred before the inferior conjunction. The apparent diameter of Venus was always significant, only once did it measure close to 35°, namely on January 24, 1806. Schröter believed that this phenomenon always occurs at a certain angle of elongation before or after the inferior conjunction, but the difference in days cited here proves that this is not the case.

The same observer says in the Astron. Jahrb. 1809, p. 165: "It—namely the phenomenon—is therefore limited only to a certain angle of elongation of the planet; and it is also possible that it may become even rarer if, perhaps, during the revolution of Venus in 23 hours and 21 minutes, one hemisphere of the planet facing us is capable of reflecting somewhat more light than the other."

When we consider how various explanations for the visibility of the dark night side of the Moon have been put forward before one could assert itself as the true one over the course of centuries, then we shall not be surprised that the one which clearly shows us the reasons for the visibility of the dark sphere of Venus has not fared any better.

The first astronomer to attempt and publish an explanation for the visibility of this phenomenon was Erasmus Reinholdt, author of the Tabulae Prutenicae. He assumed that the moon has its own special light, which becomes visible to us during lunar eclipses and in the period close to new moon, a view already proposed by the ancients. But Vitello, who first discovered astronomical refraction in the thirteenth century, had already offered an earlier explanation, which consisted in the fact that the moon consists of transparent and dark parts. Therefore, when the sun's rays penetrate those diaphanous parts that happen to be on the night side of the lunar sphere, the moon becomes visible to us.

Tycho, however, had a different opinion, attributing this visibility to Venus, which illuminated the dark disk of the moon with its bright rays. Among the ancients, one should really name Posidonius, who sought to explain this phenomenon; his hypothesis was also accepted by Oleome and, in the Middle Ages, by (Christoph) Scheiner, the discoverer of sunspots. It is based on the idea that the moon absorbs the light received from the sun into its surface and later re-radiates it. This explanation may have prevailed for a long time, for no other is known from that time until the time of the author of these tablets.

Leonardo da Vinci and Möstlin cited another explanation for this visibility, which Kepler and Gassendi also agreed with, and which is still recognized as the correct one today: that the earth's light illuminates the dark surface of the moon. Riccioli, who liked to refute other people's hypotheses, also held this opinion.

Longomontanus, however, could not be convinced by the explanation just given, for he believed that the dark lunar disk was like a concentrating mirror, which captured and re-radiated the light of the stars. We have already cited the explanation of the famous scholar of his time, Fortunatus Licetus, which, however, was never accepted as correct by any astronomer.

Cleomedes macht in seinem Werke: "Teber die Sphäre" folgende Mittheilung: Posidonius therefore says, that not only the surface of the Moon is illuminated by the Sun, like solid and opaque bodies; but to enter its thickness as thin as possible, they do not penetrate all the way, nor very deeply, because it has an enormous diameter, and the Sun itself is very far away, but the cloudless air allows the rays to penetrate easily, as it has no depth.

If so many different explanations have been attempted for a phenomenon as well-known as the one mentioned here, we should not be surprised that the causes of Venus's secondary light were not immediately discovered, and that different astronomers had their own interpretations.

For, in theory, this phenomenon presents even greater obstacles to a natural interpretation of its causes than those before which a long period of time had already accumulated a multitude of observations, all of which could be compared. However, in our case, there are very few of them, and a definitive clue is lacking here, one that, while unknown to date, nevertheless alone leads to a correct identification of the cause.

When, in 1793, as noted above, von Hahn turned to Bode and asked for an explanation of the visibility of the secondary light of Venus, he added: "I am convinced that they do not understand me as if I could attribute this dim light either to illumination from the sun or to reflected earthlight, since this explanation is completely out of the question. Unless there is some truth at the basis of this observation, only one idea remains, which certainly deserves the greatest attention."

The Berlin astronomer replied: "I suspect, namely, that the circle of dispersion of the vivid light rays of Venus is capable of producing this phenomenon." This is probably the first known explanation of this rare phenomenon, but how little it was considered correct by the scholars of the time is evident from another that Schröter gave on the occasion of his own observation.

"On the day of the local observation, February 14, 1806," says Schröter, "the apparent diameter of Venus was about 48 inches and appeared illuminated by only about 1/2 a.m. Nevertheless, in the dark room of the observatory, the rods and the lead of the window cast a very distinct shadow on the gray wallpapered wall. Therefore, it is always possible that the Earth, which appears about 50 inches in diameter on Venus, illuminated by 7/8 a.m. there, reflected some light onto the night side of Venus. However, I confess that this alone does not seem sufficient to me to account for the visibility of the dark sphere of Venus.

Dr. Herschel even leaves it undecided whether the ash-grey color of the moon's night side is not a peculiar light, and whether all planets in general do not have a faint, peculiar light. If it now turns out, it seems to be that the light from the night side of the moon is at least largely reflected by the earth, which appears four times larger there than the moon to us. Nevertheless, the comet of 1799 and the second one of 1805 have taught us that dark celestial bodies, in proportion to their mass and attractive force, attract ethereal particles to themselves up to a relative distance, which are modified into light. It is therefore more likely that both are simultaneously the cause of such a dim light. Perhaps, in the peculiar light of this planet, a random alternation of stronger and weaker light also takes place.

Harding answers the question of what makes the night side of Venus visible in the following way: "When we perceive the dark part of the moon's disk turned away from the sun, no one doubts that this phenomenon is due to reflected light from the Earth; but could the Earth also send so much light to Venus that its entire night side could still be visible from such a considerable distance? It is true that the light which Venus sends to our Earth is bright enough to make even the shadows of the opposing bodies visible, and this at a time when its hemisphere, illuminated by the sun, is only visible from Earth up to about one-eighth of its illuminated disk, as was the case on January 24, 1806.

Here Earth, then visible from Venus up to seven-eighths of its illuminated disk, should appear brighter to this planet than we saw it, even though the sunlight here is only half as strong, if the dense Earth's atmosphere were to make no changes in this regard. However, if reflected "Is it possible that, when the light from Earth caused the dark sphere of Venus to be visible, this phenomenon occurs just as often as this planet comes into a similar position relative to Earth and the sun? But even the most diligent observers of Venus tell us nothing about this, for what Count von Hahn once observed (see the previous communication) does not seem to entirely agree with the observations described above."

"I am therefore inclined to consider this dim light on the night side of Venus to be the effect of phosphorescence, according to which this planet has the power to develop so much light from its body that its hemisphere lying in the night can thereby be seen, even from such a considerable distance."

Given the current state of our knowledge, the impossibility of such phosphorescence cannot be definitively demonstrated, since we know that the property of developing light exists both in individual bodies on our Earth and, most likely, in the Earth as a whole. The bright, dim light often observed on moonless nights, in which distant objects can still be recognized with considerable clarity, is probably nothing other than such a temporary phosphorescence of our Earth.

"A few years ago, I myself ventured to explain the dim reddish light of the moon during its total eclipse partly by such phosphorescence, and I still find no reason to abandon this opinion; rather, it seems ever more probable to me, since then, Mr. Justice Councilor Schröter has also been inclined to consider the mysterious nebula and tail of comets as the product of such a phosphorescent property of these celestial bodies; a hypothesis that has received the approval of several of our best physicists and astronomers."

A more convincing and natural explanation of this rare phenomenon can be found in H.Klein's Handbook of Astronomy, p. 69. "If one does not assume," says the author, "that all parts of the Venusian atmosphere produce that light—namely, the secondary light of the night side of Venus—which is in any case improbable, but rather that the phosphorescent glow originates in the deeper parts or on the planet's surface itself, then the brightness must decrease toward the edges of the disk for our view, since we look obliquely toward those parts of the surface, and the light ray that reaches the observer's eye from there has to travel a much longer path through the Venusian atmosphere and consequently suffers a greater extinction than a ray emanating from the center."

“If, on the other hand, there were no atmosphere of Venus, the edge of the disk would appear just as bright as the center, because the radius of vision encompasses more material points in the same proportion as the obliquely emanating light appears weakened.

"But if a planet surrounded by an atmosphere is illuminated from the outside, then, unless that hazy envelope absorbs the light like a dense layer of fog, the edge of the disk appears somewhat brighter than the center. This actually occurs with the secondary light of Venus, as observers of the phenomenon saw the edge of the planet's disk stand out sharply from the background of the sky, and von Hahn, who on a few occasions saw only a small part of the disk, expressly stated that in other cases he had definitely seen the edge brighter than the center of the disk."

Thus, based solely on optical assertions, the conclusion arises that the peculiar light on the night side of Venus is probably not attributable to phosphorescence from the planet, but rather that it is the reflection of illumination from another celestial body.

Klein's claim is fully substantiated by the observations we have reported, made by Andreas Meier in Greifswalde, Schröter in Lilienthal, and Harding in Göttingen, all of whom unanimously report in their reports:

"The edge of the planet was clearly visible against the background of the sky."

"The investigations conducted by J. Rheinauer in 1859, assuming that the Earth illuminated the night side of Venus, resulted in the conclusion that this secondary light was equal to the brightness of a star of the thirteenth magnitude. This illumination, even under the most favorable circumstances, would hardly reach the magnitude of a star of the tenth magnitude, which always presupposes a very great dimness."

"The observation of the phenomenon occurred both at noon and at dusk; however, it would be possible to observe a star of the tenth magnitude by day with a 16-frame ordinary telescope, with which Kirch observed it, since even the Dorpat refractor only rarely shows a star of the ninth magnitude by day. Therefore, reflected earthlight is not the cause of the secondary light (as the observation discovered by Winnecke in Der Ham's writings proves), but rather it is produced by a moon orbiting the planet."

All of Klein's remarks are so convincing that it would be very difficult to refute them by other reasons, because they are taken from the observations themselves; they are not conjectures, such as are sometimes made to explain causes that are entirely unconscious to us, but S. Riccioli's expression of a fine ring of light on the band of the night side, p. 145.

They testify to a correct understanding of a problem whose complete solution is reserved for the near future. We therefore agree with the opinion of the author of that excellent work, all the more so since the foregoing allows us to interpret only this single possibility for the origin of the secondary light by the moon of Venus.

But these reasons become even more convincing if we go back to the position of the orbit of such a moon and its position in it at the time of the phenomenon, and try to find out whether, according to our calculations, although not yet exact, we can find such a position for this satellite, since in fact, if this latter is luminous, it can produce such a light.

Harding and Schröter observed a rare phenomenon on the surface of Venus, already mentioned, on May 7, 1799, during the transit of Mercury in front of the Sun. Schröter states: "We observed an even rarer phenomenon than that of the secondary light of Venus, namely a bright spot in the night hemisphere of this planet, which spot followed its rotation." (Astronomer. Yearbook 1812, p. 221.)

This communication by the famous Lilienthal astronomer has remained completely unknown, as it is not cited in any of the relevant works. It could not be explained by him, like similar points of light in front of the planet's illumination limit, as illuminated mountain peaks, since it occupied too large an area and was too far from that limit. This spot was the moon of Venus. To better confirm the above, we will examine phenomena that we have assumed also originated from that celestial body and were therefore produced by a superposition of it with its main planet. In all of them, the satellite would have to be located at one of its nodes on the orbit of Venus; they would therefore all contain the entire mean orbital period of 12,687 days, or half of 6,0843 days.

ROTATION TIME OF VENUS

Perhaps for no other planet has the determined rotation period shown such uncertainty and led to such lively discussion as for the planet closest to us. Every astronomer will admit that the rotation periods of the sun around its axis, as well as those of all the other planets, determined from many years of observation, are beyond doubt, but regarding the rotation period of Venus, opinions were divided. And this difference of opinion was not only a matter of the present; it became so lively at the end of the last century that the St. Petersburg Academy of Sciences made this subject a contest, without, however, leading to a definitive result. But what are the observations based on which such different determinations of this period were derived? What is the basis for the now known rotation period, and finally, the difficulty of providing a satisfactory answer to these various opinions?

It was the first scholars of their time who, as soon as telescopes were discovered, undertook to observe a planet that, as the most beautiful celestial body, had attracted the attention of all peoples. When the spots on the disks of all the planets near us were observed and their variability was discovered, curiosity arose with the task of investigating the time after which the spots, starting from a given epoch, resume their previous shape. In doing so, it was often overlooked that the spots gradually observed might belong not to the surfaces of the planets, but to their atmospheres, and do not always reappear after the celestial body has completed its rotation around its axis.

The first astronomer to make observations over a period of three years on the position of the poles and the rotation period of Venus was the often mentioned Bianchini in Rome; careful observations with the largest telescopes of his time made it possible for him to make accurate drawings of the spots to examine their various shapes and from them to deduce the time after which they returned to their earlier form, as shown to him by his illustrations.

If such a tireless observer as this scholar occupied himself for such a long period of time with solving the question of the rotation of Venus around its axis, it can be concluded that he would find a result that also agreed with his observations. Schröter's aphroditographic fragments also contain an illustration of the spots observed by Bianchini, which lay in the center of the surface. His discovered rotation period of 24 days 8 hours was later seriously doubted, until more recently it too found its defenders, who also had their reasons for accepting Bianchini's rotation period.

Alexander von Humboldt makes the following remark in his Kosmos: “Bianchini’s result has been defended by Ilussey and Flaugergues; even Hansen, whose authority is justifiably so great, considered it the more probable until 1836 (see Schuhmacher's Jahrbuch for 1837, p. 90)."

Another determination of this period, which is still considered the most accurate, but which actually still requires further testing, emerged. Dominic Cassini, a rare and highly gifted astronomer who made major discoveries in various fields of astronomy, attempted to determine the rotation period of Venus as early as 1666, but was unsuccessful. Continued observations of Venus revealed no spots for a long time, until he noticed a bright spot near the limit of illumination, which, he believed, must complete its rotation in almost 23 hours. However, as a long-time observer, he was always accustomed to protecting himself from deception and thus did not dare to assume that the bright spot he observed had always been the same over time. Since he was unable to observe the progression of the bright spot over a sufficiently large arc, he left the question of the rotation period undecided.

His son, Jaques Cassini, however, believed that he had determined from his observations that the period found by Bianchini was 25 times longer than the actual period. He published in the Memoires de l'Academie of 1732 and in his Elements of Astronomy the period of 23 hours 22 minutes derived from his observations. At the same time, he made the remark that the spots which Bianchini had observed at different times were probably different from his own, which then explains the erroneous result which he had found.

When Schröter began his observations of Venus in Lilienthal, he first sought to familiarize himself with the shape of the rarely appearing spots. He also wanted to equip himself with highly magnifying telescopes to make observations that could provide new insights into the physical constitution of the celestial bodies.

After continuing his painstaking observations of the surface of Venus for several years, he made the following remark: "I soon discovered from the continued observations of the unequal horn tips of Venus that Bi's assumption was also unfounded; while Cassini's statements, on the other hand, are more in agreement with them, the shape of the southern horn, as well as that of the northern one, changed in just a few hours and coincidentally returned after 23 to 24 hours, and from this it followed not only this period of Venus's orbit, but also that its equator is considerably tilted toward the ecliptic and its poles must be quite far from the horn tips."

Schröter, as is well known, considered the points of light sometimes observed on the horns to be mountain peaks on the surface of the planet and sought to determine the rotation period of Venus from the alternation and recurrence of the same shapes of these horns. From his observations from 5 p.m. on December 28, 1789, to 10 a.m. on December 25, 1791, i.e., in 726 days and 17 hours, he found an average period of 23 hours, 20 minutes, 59.4 seconds. The aforementioned period contains 747 such periods, with only a very small difference of 7.1 of such a period.

The tireless observer chose an even longer period, namely from 5 p.m. on December 28, 1789, to 5:40 p.m. on February 27, 1793, which therefore amounted to 1157 days. 1 hour 41 minutes and found that this contains 1189-28 periods. Through these tests, Schröter convinced himself of the correctness of his period, which gave him a second result for the rotation period of Venus 23 hours 21' 19", or 20" longer than the previous one. He sought this last determination from the shape of the tips of the horns always returned after 23 to 24 hours does not always seem to follow from the observations in the Aphrodit. (Astron. Jalirb. 1795, p. 211.

To derive the equation of the blunt horn with the pointed one. Long after the publication of this determination, namely in 1809, the Lilienthal observer undertook a further revision of his observations, resulting in a final result of 23 hours, 21 minutes, and 7.97 seconds. Schröter concludes his communication of this rotation period he discovered with the words: "It can now be asserted with conviction that the rotation period of the planet Venus has been determined as reliably as was ever possible given the extraordinarily mountainous nature of its surface."

Fritsch sought a close confirmation of this period in Quedlinburg by comparing the illumination limits, by a bending in the northern part of the same; after three days of observations, he believed that the rotation period of Venus was 23 hours 22 minutes (Astron. Jahrb. 1804, p. 214).

While Venus in 1836 occupied a position in the sky which, especially in our region, could be described by observers as a very advantageous one, Beer and Mädler in Berlin also made a series of observations of this planet. They confirmed what Melireres had observed by Schröter, but with regard to the rotation period, they believed several times that the same shape of the horns would have been observed after 23 to return in 24 hours.

Schröter's above-mentioned time of rotation was confirmed by de Vico, the director of the Roman Observatory, since he clearly recognized spots on Venus from 1840 to 1842 and, with the help of an excellent achromatic telescope from Gaucho ix, was able to determine the rotation time to be 23 hours, 21 minutes, 21 to 93 seconds.

The mean period previously determined from the various phenomena of Venus's satellite appears to be very closely related to the rotation period of the planet found by Bianchini, for it is only half of it, with one insignificant difference. If one takes the arithmetic mean of the five mean periods given from pages 104 to 138,

I. 12-1708

II. 12-1708

III. 12-1682

IV. 12-1679

V. 12-1666

This gives us 12-1687 days for the satellite's orbital period. Double this period is 24-3374 days; the rotation period determined by Bianchini is 24-3333, so the difference is 0.0041 days.

This small numerical value of 6 minutes 3 seconds, by which these two distances, which were found by such different methods, differ, leads us to believe that Bianchini had at that time observed the satellite of Venus on her disk, which he regarded as an ordinary spot.

This, however, leads us to the conviction of Jacob Cassini, who had found a different rotation period, namely, "that the spots observed by Bianchini were probably different from those of his observations." However, this very similarity between our found mean period and the rotation period of the Roman scholar entitles us to the conclusion we have already expressed, that Bianchini did not observe random spots, but rather the moon of Venus, which he only ever saw after two orbits on the disk of Venus.

This rare observation can only be explained by the fact that the companion of this planet only completes its own rotation period after two orbits around the main planet. To the persistent observer, the satellite only became visible when it appeared to him as a spot on Venus, since it then turned its dark side toward us, but as a bright point of light or bright spot, it became invisible on the main planet, since its albedo, the intensity of its light, is no different from or equal to that of Venus. Only in rare cases, such as when the main planet is very close to us, does it become visible as a striking point of light.

A second application of the rotation period discovered by Bianchini leads us to the direct derivation of the results found by Schröter and de Vico from it, but it is not easy to explain how this connection between them arises. According to Cassini's opinion, already reported on page 166, the actual rotation period of Venus is 25 times shorter than that of Bianchini; thus, assuming it to be 24-3333 days, 1/ro of it is equal to 23 hours 21 minutes 35-71 seconds.

Now Schröter's 1st rotation time 23 hours 20 minutes 59.4 seconds, difference from Y 25 Bianchini = -f- 36-31", by which the latter is greater, Schröter's 2nd rotation time 23 hours, 21 minutes 19 seconds, difference from V 25 Bianchini = + 16-71".

Schröter's 3rd rotation time 23 hours 21 minutes 7 '97 seconds, difference from Bianchini = + 27-74".Furthermore, De Vico's rotation time 23 hours 21 minutes 21-93 seconds, difference from V25 Bianchini = 13-78".

The latest and most reliable rotation period of de Vico differs from the 14.5 rotation period of Bianchini only by 16", of which Schröter's only in the first case by twice this amount.

However, the fact that Schröter's result can also be derived from that older determination is probably a consequence of the time intervals he chose, which always contain exactly the orbital period of Venus's satellite determined by us, i.e., half of Bianchini's rotation period. Such a derivation will confirm what has just been stated.

1) From 2 April at 10:15 p.m. in the evening, 1793, to 2 August at 12:40 p.m. in the afternoon of the same year, 121 days, 14 hours, 35 minutes have passed. According to Schröter's rotation period, this time amounts to 125 revolutions, up to an insignificant fraction 001, but according to Bianchini's period, it amounts to 5 of the same, leaving a fraction of 0-008.

2) From 26 February to 14 August 1793 are 168 days, which gives 7 revolutions compared with Bianchini's and there remains a surplus of 0.38.

3) Since March 13, 1793, at 11 a.m., when an apparition appeared on the horn, which was seen again on August 2, in the evening at 12:40 p.m. of the same year,

142 days, 1 hour, 40 minutes have passed; this time includes six of Bianchini's periods with an error of 0-64.

The above observations are all taken from Aplirodit, Fragm. Schröter's; there is no doubt about the correctness of the famous author's derivation, but I believe that I have proven with these few examples what I have stated before, that from these observations Bianchini's period, as well as half of it, namely ours, can be derived.

It is well known, however, that Schröter derived his rotation period from the variation in the southern horn of Venus, but Bianchini did not use this in his determination, as he was not yet aware of this striking variation in its shape.

I do not know whether de Vico derived his rotation period from other phenomena than the one mentioned above and observed by Schröter; it can be assumed that he proceeded in the same way as his predecessor in Rome, only he probably observed other spots.

The above account of Bianchini's precise observations prompts me to mention another no less valuable determination by him, namely, the elevation of the pole of Venus, discovered by him, of 15° above the plane of the ecliptic. According to this calculation, the equatorial plane of the same planet should rise above that plane by 75°, and the spots he observed should move at this inclination to the ecliptic.

If, then, the opinion we expressed earlier, that the same scholar had observed not the spots of this planet, but its moon, is accepted as consistent with the nature of the matter, it would follow that the inclination of Venus's satellite was 75°. From Lambert's determinations, which we have already communicated, based on observations of the same satellite by Montaigne, the inclination of that moon of Venus is 64°.

I have repeated the same calculation and found it to be 63° 40', which is the same result that the worthy scholar published 100 years ago. Consequently, there is only a small difference of 11° between Bianchini's and Lambert's determinations. Thus, here too, a proof for the existence of the moon of Venus is shown, borrowed from completely different observations.

Comparison of the observations with the assumption of the known orbital period. In the last century, no one has attempted to determine the orbit or the orbital period of the moon Venus as tirelessly busy, like Lambert, for he sought its location for the corresponding time from the available information about its perception, so that he perceived a certain regularity in its movement and thus convinced himself of its existence.

The average period or orbital time of the moon of Venus, as determined by us, is 12 days and 3 hours. It travels approximately 30° in its orbit.

It is noteworthy that the first observation agrees exactly with our daily motion. Furthermore, the largest error in this comparison is up to -35°, which, in time terms, is the difference between the mean motion and the true one.

Thus, what Lambert presented as proof for the existence of the celestial body, based on the observations of the satellite of Venus known to us, is presented here. He says: "But in addition to the means proposed by Mr. Hell for detecting optical deception, there is another one that directly concerns the satellite itself. For if a satellite is present around Venus, then its phenomena must also appear as a satellite, and therefore its motion must be lawful and orderly."

The mean distance of the moon of Venus from its main planet

It is known that it is possible to determine the mean distance of a planet surrounded by satellites from the sun from the mass ratio of the sun to that of the planet itself, provided the mass, distance from the sun, and the orbital periods, namely those of the main planet around the sun and those of the satellites around their own planets, are known. Newton and Laplace have given the relevant formulas, from which, assuming the mass of Venus is known, the mean distance of the moon of Venus from the center of its main planet can be determined.

This distance is only slightly less than that of our satellite, from which it differs in many respects. The eccentricity was determined from the observations of 1761 and is therefore significantly larger than that of any large planet or satellite; it is half the size of Mercury.

The determination of the inclination is possible from these same observations.

From the review of the previously communicated observations it appears that, according to those of Cassini in 1672, the moon in question must have been near the node of its orbit; therefore, the ascending node of the moon for 1672 appears to be 2 Z. 20°.

Lambert, however, found almost the same numerical values from Montaigne's observations, so there is no great difference here. The aforementioned scholar sought this derivation from the four observations of that year, namely from the reported distances of the Moon from the main planet.

The other determining elements of the orbit cannot be determined from the present observations, but should it hopefully be possible in the future to obtain more determinations of its location, these will also be found and will hopefully surpass in accuracy those which can only claim to be an approximation from inaccurate location information.

Presence of dark bodies in the universe

No science offers such a long, continuous series of discoveries as astronomy, which, through the research of rare, brilliant scholars, has now assumed considerable scope in all its branches. In a period of barely a hundred years, our view of our solar system has expanded so significantly that we now look into the future with a free, unprejudiced gaze, expecting from it the most magnificent insights into those unknown regions of the solar system, both near and far, with its smaller and larger celestial bodies.

It was undoubtedly the famous French mathematician La Place, who first proposed, in his work Exposition du Systeme du Monde at the end of the last century, that there might be as many dark celestial bodies in the sky, invisible to us, as there are bright ones. In a treatise devoted exclusively to this subject, he proved, with the help of advanced analysis, that celestial bodies of very large masses can exist that could be said to emit no light, and are therefore dark and invisible. The actual impetus for this scientific work was the explanation of the sudden appearance of shine and disappearance of the star discovered by Tycho de Brahe in 1572 in the constellation of Cassiopeia, which appeared on November 11 and remained visible until the beginning of spring 1574. A number of respected naturalists of the time, such as Theodor von Beza, Reisaclier, Hieronimus Fracastor, Johann Den, Elias Camerarius, and Peucer, attempted to interpret the appearance of the mysterious star through various opinions.

This also prompted the discoverer to publish a short work, which constitutes the first part of his Astronomiae instauratae Progymnasmata, Pragae IG 10, in which he presents the hypotheses of the aforementioned scholars.

What a difficult task, then, if Laplace, by his rare insight, discovered a hitherto unknown truth, the future explorers of the heavens will have to discover in the various spaces of the universe those dark celestial bodies which no telescope, no uninterrupted search of the tireless astronomer, can find, but whose existence can only be recognized by the proximity of the bright celestial bodies which form a system with them.

If we assume only one star for every square minute of the firmament, whose total surface area encompasses 41,252 square degrees—which is a low assumption, especially for the regions of the Milky Way—we arrive at a total of 148,506,500 fixed stars throughout the entire celestial sphere. If, of these millions of stars observable with our present means, not much more than 100,000 have been observed, this sufficiently defines the task of astronomy for the future. But if, according to Laplace, there were just as many dark celestial bodies, the standpoint of contemporary science would by no means be such that we could believe that the greatest discoveries in its field could already be considered to have been made.

The assumption of dark bodies in the universe gives rise to a special branch of astronomy, namely, "the astronomy of the invisible." What Laplace expressed as an opinion was proven by Bessel, who, guided by careful observations of the motions of Sirius and Procyon, expressed the firm conviction, even in the last years of his career, that a dark secondary star must accompany Sirius and cause its peculiar motion.

The famous Königsberg astronomer, who is known to have been the first to determine the parallax, or distance, of a fixed star, has enriched astronomy in modern times with his finest scientific works. He was the first to introduce such great precision in both calculation and observation that this can be considered his outstanding achievement, for in this he surpasses all his predecessors.

Through the most careful testing of the accuracy of his observational tools, he discovered the particular errors inherent in every astronomical instrument, which, by constantly taking these into account, enabled him to provide almost error-free observations. How necessary this care is, however, in the observation of fixed stars, is demonstrated by the small changes discovered in the positions of each star, continuously observed for a number of years. He tried every possible method of observation, which he extended to every subject of practical astronomy.

Through these admirable and careful observations of all heavenly bodies, the tireless astronomer deserves a name that will be preserved until the very future.

Bessel was not to live to see the brilliant confirmation of his firm conviction, which was to become an acknowledged truth soon after his death by Peters, Mädler and Schubert in their excellent investigations of some fixed stars.

Peters, then in Königsberg in 1851, was attempting to determine, based on Bessel's hypothesis, that Sirius formed a system with a second, invisible, dark mass, and that both, according to Newton's laws of attraction, revolved around their common center of gravity. He was attempting to determine the previously observed inequalities in the motion of this large fixed star, and from this, to determine Sirius's own elliptical orbit. The success of his efforts fully met expectations, for an orbit with an orbital period of 50.9

years represented such a uniform apparent orbit that comparing it with the previously noted deviations in the position of the fixed star made it disappear, except for the smallest differences. Mädler undertook similar work in Dorpat almost simultaneously with Peters on the fixed star Procyon, which ultimately allowed an orbital determination corresponding to an orbital period of 50 to 60 years.

Somewhat later, a similar scientific work on a Virgo or Spica was published by Schubert in Cambridge (United States), which revealed an orbit with a 40-year orbital period.

These investigations have established the existence of dark celestial bodies in the universe, which, particularly through their determined orbital periods, clearly point to the planets of our solar system, from which they differ only in their greater mass and lack of any illumination. If we assume the existence of dark acolytes everywhere in the universe, including our solar system, the following classes of bodies would likely be found.

We will try to name for each of them well-known phenomena, some of which have already been recognized as striking.

1. A bright fixed star and a dark one form a system, the latter rotating around the first.The three stars just mentioned, Sirius, Procyon, and Spica, belong here, and will be followed by investigations of others. However, all or most of the variable stars, which differ from the first-mentioned in that their orbital planes lie in our extended line of sight, should probably be included here. Since these known fixed stars are almost all smaller than first-magnitude stars, they have been observed less frequently, and no or only a very slight deviation in their position has been observed.

2. Fixed stars with a partly dark surface, which, due to their slow rotation around their axis, periodically and after a long time, show us their light side
turn to.

One might include here the stars known as new or vanished, such as those of 1572, 1604, 1670 and others.

3. The aerolites, which have only been observed frequently in our century, and which have sometimes been observed in large numbers even for short periods beyond the boundaries of our solar system, can be considered as such. Although, through the efforts and discoveries of Schiaparelli, Erman, Newcomb, Leverrier, and von

Oppolzer, it is now accepted as a fact that these small bodies are components of cometary tails, which are only visible for a short time as they approach the Earth, this only applies to periodic meteorites, such as those of August 10 and November 13; the so-called sporadic ones, however, appear almost daily and give us an idea of how many of them must be moving in the solar system as small, dark bodies. Outside our solar system, in 1796, Schröter observed a group of aerolites momentarily passing through the field of his reflecting telescope in the constellation of Opiuchus.

4. A dark satellite orbits a bright planet in our solar system. This includes the moon of Venus discussed in this paper, whose existence has been doubted for so long. The three fixed stars just mentioned, Sirius, Procyon, and Spica, provide us with the information that their dark companions do not receive light, but rather absorb it, so to speak. For, since they are hardly more distant from the large fixed stars than Saturn is from our sun, they would otherwise reflect the light falling on them and become visible to us. In general, the illumination of the planets in our solar system, including those of the known emanation or vibration hypothesis of light, does not always seem to find its

application, especially with regard to the most distant upper celestial bodies.

On the two planets Uranus and Neptune, the sun appears to have such a small diameter that one cannot possibly conclude that its bright illumination comes solely from the sunlight. If, according to the well-known so-called Bode progression, we were to determine the distance of the planet located after Neptune, if such a planet still existed, we would obtain a size for the visible diameter of the sun on that planet that would hardly be visible to the naked eye in the sky. It is known that the discovery of Uranus by the elderly Herschel on March 13, 1781, was caused by an error in Wollaston's star catalog.

Bode soon discovered that this planet had already been observed as a fixed star in 1690 by Flamstead and in 1756 by J. Mayer. Likewise, soon after the discovery of Neptune, Walker in Washington discovered two locations of the same, dated May 8 and 10, 1795, which were recorded as fixed stars by Lai an de in his famous work *Ilistoire celeste*. They are found there on pages 156 and 158, and although Lalande always adds remarks about the intensity of the light or the colors of other stars, these are missing in these two observations; therefore, the planet differed in no way from any other fixed star. In the following small table, we present the various degrees of illumination of the celestial bodies of our planetary system.

The celestial bodies of our planetary system.
Apparent luminous intensity,

The comparison of luminous intensity reveals such a low light intensity for the upper planets, especially Uranus and Neptune, that we must ask ourselves whether they derive the great brightness they appear to us as fixed stars solely from the sun, which there appears under a small diameter? However, not only these planets, but also their moons appear to us with such a strong light that they too resemble very small fixed stars.

This different illumination of nearby and distant planets by sunlight cannot therefore be fully explained by either explanation for light and its motion; it appears to be caused by an interaction between planets and sun, so that the celestial bodies further away from the latter and in need of its illumination nevertheless develop as much, perhaps partially, as they need.

The above investigations into the satellite of Venus, from the various reported phenomena, which could not be explained without applying certain assumptions, undeniably indicate that they all belong to one and the same celestial body, whose existence has so far been doubted. Only the assumption of the presence of such a body offers a sufficient explanation for them, for if there is no moon around Venus, then these phenomena can also be regarded as mere coincidences.

However, they are also interrelated, and one can be deduced from the other. However, if the existence of a companion of Venus is proven in this way, one can still counter that it is not visible, and indeed, it has not been observed for more than a hundred years. We repeat here once again the opinion of two great and worthy scholars, Cassini and Lambert, on this matter. The first stated: "The newly discovered moon of Venus rarely reflects enough light to be seen."

The other remarks: "In fact, one sees no necessity for all planets and satellites to be brilliantly bright or to constantly shine with full light. After all, there are fixed stars that are only occasionally visible."

These opinions have also been confirmed in recent times. The existence of such a star has been put beyond doubt by the dark acolyte of Sirius. What was the famous star of 1572, discovered by Tycho, and why is it no longer visible? What were similar fixed stars that suddenly appeared and disappeared, and also became completely invisible?

Finally, in the starry sky of the south, we find two dark spots known as the "cabbage sacks." They appear to us as devoid of brightness and light only because the parts of the Milky Way immediately adjacent to them are extraordinarily brilliant. Couldn't the brightly shining Venus, in the present case, also obscure its moon with its rays, making it invisible to us?

But if there are dark fixed stars, then dark secondary planets may also exist, one of which is the moon of Venus. This leads us directly to the explanation of variable stars, whose light is dimmed by the passage of a dark companion. The two upcoming transits of Venus across the sun will, if the satellite happens to be near Venus, provide the best observations to prove its existence. If six different phenomena can be properly explained by assuming that they are produced by a celestial body unknown to us as yet, then the existence of the same can be assumed with the greatest probability.

BIBLIOGRAPHY

Writings about the moon of Venus

Bautlouin, On the Discovered Satellite of Venus. Berlin 1761, 8.

Baldwin, On the Satellite of Venus. Paris 1761.

M. Hell, On the satellite of Venus. Vienna 1762

J. C. Lambert, The Satellite of Venus (Astron. Jahrb. 1777. Berlin).

History of the French Academy for 1741

Philos. Transact. Nr. 459.

Encyclopedie. Tom. XVII, p. 837.

CORRECTIONS.

Page 5, line 17 from the top reads: Jahnsen instead of Johnsen.

Page 5, line 6 from the bottom reads: eccentric scholar instead of unlearned charlatan.

Page 7, lines 14, 17, 27 from the top reads: Jahnsen instead of Johnsen.

ENDE

www.ingramcontent.com/pod-product-compliance
Ingram Content Group UK Ltd.
Pitfield, Milton Keynes, MK11 3LW, UK
UKHW020143250726
13967UKWH00002B/840

9 780930 472900